APPC N

Improving Chemical
Engineering Practices

Forethoughts

"It ain't so much the things we don't know that get us in trouble. It's the things we know that ain't so."

Artemus Ward (1834–1867)

"Cultures are slow to die; when they do, they bequeath large deposits of custom and value to their successors, and sometimes they survive long after their more self-conscious members suppose them to have vanished."

Irving Howe, *World of our Fathers,* Simon and Schuster, New York, 1976, p 618

"Time strips our illusions of their hue,
And one by one in turn, some grand mistake
Casts off its bright skin yearly like the snake."

Byron (1778–1824), *Don Juan,* 5.21

"Superstitions take a long time to die—almost as long as new ideas take to be born."

W.S. Sykes, *Essays on the First Hundred Years of Anaesthaesia,* Churchill Livingstone, Edinburgh, 1962, p 90

"It is important that students bring a certain ragamuffin, barefoot irreverence to their studies; they are not here to worship what is known, but to question it."

J. Bronowski, *The Ascent of Man,* BBC Publications, London, 1975

Improving Chemical Engineering Practices

A New Look at Old Myths of the Chemical Industry

Second Edition

Trevor A. Kletz
University of Technology
Loughborough, England

● HEMISPHERE PUBLISHING CORPORATION
A Member of the Taylor & Francis Group
New York   Washington   Philadelphia   London

Improving Chemical Engineering Practices: A New Look at Old Myths of the Chemical Industry

1 2 3 4 5 6 7 8 9 0   BBBB   8 9 8 7 6 5 4 3 2 1 0 9

This book was set in Century Schoolbook by Bi-Comp, Incorporated. The production supervisor was Bermedica Production; and the cover design was by Debra Eubanks Riffe.
Braun-Brumfield, Inc. was printer and binder.

Library of Congress Cataloging-in-Publication Data

Kletz, Trevor A.
    Improving chemical engineering practices : a new look at old myths of the chemical industry / by Trevor a. Kletz. -- 2nd ed.
        p.   cm.
    ISBN 0-89116-929-6
    1. Chemical engineering.   I. Title.
TP155.K544   1989
660--dc20   89-19814
                                                                      CIP

Contents

Preface

During my years in the chemical industry, as a manager and particularly as a safety adviser, I came to realize that many chemical engineers accepted uncritically a number of statements of doubtful accuracy. I therefore thought that it might be useful to set down these myths, as I have called them, and to say why I think they are wrong, partly because they have led to some accidents and wrong decisions and partly to encourage a sceptical approach, particularly among younger engineers. I hope, therefore, that the book will prove useful to teachers and students as well as to those in industry.

For this U.S. edition I have added another 16 myths, making a total of 60 and I have replaced some words and phrases by their U.S. equivalents. For example, I have changed *road tanker* to *tank truck* and *manager* to *supervisor*. (In the United Kingdom *supervisor* is another name for a foreman.) There is a glossary of U.S.-U.K. chemical engineering terms in my book, *What Went Wrong?—Case Histories of Process Plant Disasters* (Gulf Publishing Co, 1985).

I am sure there are other myths besides those I have described and I shall be grateful for any contributions that readers can send me.

Thanks are due to the many colleagues, past and present, who have suggested ideas for this book or commented on the draft, particularly Professor F.P. Lees and Mr. K. Palmer; and to the Science and Engineering Research Council and the Leverhulme Trust for financial support.

About the Author

Trevor Kletz retired from the chemical industry in 1982 after 38 years service; for the last 14 he was safety adviser to ICI Petrochemicals Division and was particularly concerned with process safety and loss prevention. He then joined the staff of the Department of Chemical Engineering at Loughborough University of Technology.

Introduction

Many people believe that scientists and technologists are entirely rational in their beliefs and actions, at least when acting in a professional capacity. Ordinary, uneducated people may believe in myths and act accordingly, but not engineers. A historian of the 18th century has written:

> . . . by accustoming landowners and merchants to think in objective terms about how their resources might be employed most profitably, the ethos of capitalism led them to devalue (although not abandon) purely emotional criteria in evaluating men and measures in other spheres. Learning to think in a hard-headed way about profits helped to relax the hold of older irrational feelings as guiding principles in public policy.[1]

Though scientists and technologists may be less prone to a belief in myths than other people, they have not totally abandoned them, as I shall show. Many half-truths are half-believed and they are not mere whimsies, like a belief in fairies. They affect our actions. I shall describe many accidents and wrong decisions that were, in part, caused by a belief in myths.

1

Myths have several characteristics:

1. They are not completely untrue; there is usually a measure of truth in them, but they are not completely or literally true either.
2. They were often more true in the past than they are now.
3. They are deeply ingrained. When our reasons for believing in them are shown to be invalid, we look for other reasons, or continue to act as if the myths were still true. Once in the mind, they are there to stay.

Readers may like to consider the extent to which these features of myths apply in other walks of life. Here we shall be concerned only with some myths of the process industries, particularly the chemical industry, deeply ingrained beliefs that are not wholly true. Few, if any, people believe all the myths listed below; most believe—or half-believe—some of them. This is therefore a work of iconoclasm; an attempt to destroy some myths and to encourage a sceptical approach. Too many engineers seem to accept the received wisdom and see themselves as practitioners of established techniques. Perhaps I can sow seeds of doubt. Most of what we have learned is good, sound stuff, but there is some chaff among the wheat.

A final feature of myths is that often we take them for granted and never seriously think about them. Once we write them down, as in the headings to the following sections, we begin to have doubts.

In the following pages the myths are divided into myths about technology and myths about management. Many of them previously appeared in references 2 through 6 and have a bias toward loss prevention and process safety.

References

1. Endelman, T.E. 1979. *The Jews of Georgian England.* p. 35. Philadelphia, PA: Jewish Publication Society of America.
2. Kletz, T.A. 1974. In *Loss Prevention and Safety Promotion in the Process Industries,* Proceedings of the First International Sym-

posium on Loss Prevention and Safety Promotion in the Process Industries. p. 309. Amsterdam: Elsevier
3. Kletz, T.A. 1976. *J Hazardous Materials* 1(2):165.
4. Kletz, T.A. 1977. *J Hazardous Materials* 2(1):1.
5. Kletz, T.A. 1986. *Chem Engineer* No 431, December, p. 29.
6. Kletz, T.A. 1987. *Chem Engineer* No 443, December, p. 44.

1

Myths About Technology

MYTH 1

Pressure vessels must be fitted with relief valves (or bursting discs).

Many people are under the impression that, by law, all pressure vessels must be fitted with relief valves (or bursting discs), and in some countries this is so. In the United Kingdom, however, the law requires only certain pressure vessels to be fitted with relief devices: the Factories Act (Sections 32, 35, and 36) requires steam boilers and receivers and air receivers to be fitted with relief valves, and the Chemical Works Regulations (1922) require certain other vessels to be fitted with them. These Regulations, however, apply to only a small proportion of the chemical industry, to those processes listed in the Regulations, all of which were in use in 1922. In some countries the laws are more restrictive, and in others, including the United States, less so. Nevertheless, whatever the law, for many years it has been the universal practice to fit relief valves on all pressure vessels, and it is usually taken for granted that they will be fitted. They are required by the recognized pressure vessel codes, and it is often assumed that they are necessary to fulfil the requirement of the Factories Act (Section 176) that equipment shall be operated and maintained "in an efficient state, in an efficient working order and in good repair" and the requirement of the Health and Safety at Work Act (Section 6) that any article for use at work must, so far as is reasonably practicable, be "so designed and constructed as to be safe and without risks to health."

However, as plant and equipment get bigger, the cost of providing relief valves rises. The valves are not so expensive in themselves but, if flammable materials are being handled, the associated blowdown and flare systems are expensive and give rise to complaints about the noise and light. If toxic materials are being handled the relief valve discharge may have to pass through a scrubbing system.

Is it possible to use, instead of a relief valve, a pressure switch that will detect a rise in pressure and isolate the source of pressure (referred to later as an instrumented protective system or trip)? For example, on a distillation column the relief valve is usually sized on the assumption that the full heat input to the base continues when reflux is lost. Could the rise in pressure be used to isolate the source of heat? If so, only a much smaller relief valve will be needed to cope with other sources of overpressure. If a single pressure switch and motor valve are considered insufficiently reliable, then they can be duplicated.

The reaction of many people is that this suggestion is unacceptable because 'instruments are unreliable.' "Fit an instrument if you wish," many engineers will say, "so that the relief valve does not lift so often, but let me fit a relief valve as a 'last resort'."

The answer to this comment is that although instruments are not 100% reliable, relief valves are not 100% reliable either and it is possible to design an instrument system with a reliability as good as or better than that of a relief valve.[1,2] Table 1.1 shows typical figures for a trip system and a relief

TABLE 1.1 Hazard rates for various protective systems

Protective system	*Hazard rate* (i.e., the frequency with which the design pressure is exceeded)
No trip or relief valve	Once a year (assumed)
Simple trip,	
annual testing	Once in 5 years
monthly testing	Once in 48 years
weekly testing	Once in 200 years
Relief valve,	
annual testing	Once in 200 years
Duplicated trip,	
monthly testing	Once in 1728 years

valve assuming a demand rate of once a year (that is, if there was no relief valve or trip the vessel would be over pressured once per year) and assuming that the trip develops a fail-danger fault once in 2 years and a relief valve develops a fail-danger fault once in 100 years.[3]

The duplicated trip is actually *safer* than a relief valve and, if cheaper, could be used instead, provided that the staff concerned understand the need for regular testing and have the necessary facilities. In fact, a fully duplicated trip system is much more reliable than a relief valve and complete duplication is unnecessary. Because of the uncertainties in the figures and differences in the mode of failure (a relief valve that fails to operate at the set pressure may operate at a higher pressure, but this is not true of a trip) I suggest that a trip system used instead of a relief valve should have a reliability 10 times greater than the figures quoted above for relief valves.

What about vessels that must be fitted with relief valves by law? Can protective systems be used instead?

As far as steam boilers and air receivers are concerned the law in the United Kingdom is absolute—relief valves must be fitted; for steam receivers the law is less definite. However, we would not often want to fit protective devices instead of relief valves because steam and air can be discharged safely into the atmosphere. As far as vessels covered by the Chemical Works

Regulations are concerned, Regulation 5 states that a relief valve must be fitted to "Every still and every closed vessel . . . in which the pressure is liable to rise to a dangerous degree." It could be argued that if a reliable trip is installed, then the pressure is not liable to rise to a dangerous degree.

Similar arguments can be applied to the pressure vessel codes. The Appendix to BS 5500 states:

J.1.1   Every vessel shall be protected from excessive pressure or vacuum . . . except as provided for in J.1.2.

J.1.2   When the source of pressure (or temperature) is external to the vessel and is under such positive control that the pressure (or temperature) cannot exceed the design pressure (or temperature), a pressure (or temperature) protective device need not be provided.

A reliable trip, properly maintained, keeps the pressure under positive control.

The predecessors of BS 5500 (BS 1500 and BS 1515) contained similar wording (though without the references to temperature).

The corresponding US code (ASME Boiler and Pressure Vessel Code, Section VIII) does not contain similar wording and requires all pressure vessels to be fitted with pressure relief devices (§ UG–125). However, it seems that the letter of the Code can be satisfied if a small pressure relief device is fitted and a reliable trip is installed to keep the pressure under control. § UG–133 states that:

"The aggregate capacity of the pressure relieving devices . . . shall be sufficient to carry off the maximum quantity that can be generated or supplied to the attached equipment." Presumably, in calculating the "maximum quantity," allowance can be made for the effects of a reliable trip system.

1.1   Examples of the Use of Protective Systems in Place of Relief Valves

The following are some examples of situations in which instrumented protective systems have been, or might be, used in place of relief valves.

1.   It is sometimes possible for a compressor, handling flammable gas, to suck a vacuum in the suction

catchpot. Relief valves are sometimes installed to prevent this from occurring, but sucking air into the vessel with a consequent risk of explosion could be as dangerous as collapsing the vessel. On some systems of this type, protective systems have therefore been installed to detect a low pressure, trip the compressor, and close a valve in the suction line before the pressure gets too low. The systems have been designed so that failure will coincide with a demand and the vessel will be underpressured once in 2000 years.

2. On a distillation column, as already stated, the relief valve is usually sized on the assumption that heat input to the reboiler continues when reflux or cooling have been lost. This usually calls for a large relief valve. On a few distillation columns a high-pressure trip has been used to isolate the steam to the reboiler or isolate the fuel to the reboiler furnace. A small relief valve can then be installed to cope with excess pressure generated in other ways. Care must be taken that the residual heat in a furnace or reboiler is not sufficient to overpressure the column.

3. As part of an uprating project a scrubbing column had to be fitted with an additional steam reboiler and feed vaporizer. A larger relief valve would normally be required. The flare stack and flare header were not able to take any additional load, and their replacement by a larger system would have been very expensive. A trip system was therefore designed to isolate the steam to the reboiler and shut down the hot water pump on the feed vaporizer when the pressure approached the set pressure of the relief valve. The relief valve was left in position to cover fire relief and other requirements.[2]

4. Flammable gas at high pressure in a pipeline often has to be let down to a lower pressure. Normally, a relief valve is fitted on the low-pressure gas main to prevent it from being overpressured if the reducing valve fails in the open position. A very large relief valve is required. If there is no convenient flarestack available and no suitable site for one, a high-reliabil-

ity trip system can be installed to isolate the high-pressure main if the reducing valve fails and the downstream pressure rises.

5. Although protective systems would not normally be used in place of relief valves on steam duty, because steam can be discharged safely into the atmosphere, large condensing turbines are a special case, because they need very large relief valves to cover accidental isolation of the condensate lines or loss of cooling water. Failure to provide these valves has caused many accidents.[4] Protective systems that isolate the steam when the pressure in the turbine rises should therefore be considered.

6. In a particular plant, hydraulic overpressure of a vessel is prevented by a trip system that isolates the feed. In addition, several alarms have to be ignored before there is any demand on the trip system.

7. In oxidation plants air (or oxygen) and hydrocarbons have to be mixed and sometimes the plants have to operate close to the explosive limit. Installation of a relief device large enough and sufficiently quick acting to prevent an explosion from occurring is usually impracticable. Some companies, therefore, install blast walls around the equipment. These do not prevent damage to the reactors but merely minimize the consequences to people and other equipment. (Sometimes they do not even do that; see Myth 18.) Stewart[5] has described an oxidation plant that operates very close to the explosive limit and is fitted with a protective system of exceptional reliability similar to those installed on nuclear power stations. Fifty trip initiators, all of them triplicated, detect the approach of dangerous conditions and isolate the oxygen supply by means of a high-reliability shutdown system.

1.2   Compromise Solutions

If, despite the foregoing, you still feel that the pull of tradition is too strong, then a number of compromise solutions are possible.

1. Fit a full-size relief valve but let it discharge into the atmosphere, though this would not normally be allowed, and rely on a trip system to prevent it lifting, or, to be more precise, to make the chance of it lifting very low. Each case must be considered on its merits, but in many cases relief valves discharging flammable or toxic gas into the atmosphere should not lift more than once in several hundred years, and relief valves discharging hazardous liquids into the atmosphere should not lift more than once in several thousand years.

2. When several relief valves discharge into the same flare header the flare header can be sized to take the flow from the largest relief valve (or largest pair) and protective systems can be used to make sure that the probability of more than one (or two) relief valves lifting at the same time is acceptably small. For example, on one plant a number of distillation column relief valves discharge into a flare header that is sized to take the rate from the largest only. All the columns could be overpressured simultaneously if there was a power failure or cooling water failure, but instrumented protective systems isolate the heat supply to the base of each column and ensure that the chance of more than one relief valve lifting is small.

3. Occasionally one hears of a vessel in which the pressure is controlled by an instrumented protective system but an undersized relief valve is installed as a sort of token observance of the myth. There is no sense in this (though it may be sensible to install a trip, to avoid the use of a large relief valve, and a small relief valve to guard against other sources of overpressure).

1.3 Stronger Vessels

Another way of avoiding the use of relief valves is to use stronger vessels. If a distillation column can withstand the maximum pressure developed when the reflux is lost but heat input to the base continues, then a relief valve is unnecessary [Figure 1. 1(c)]. This may not be economic on a large column but may be on a small one.

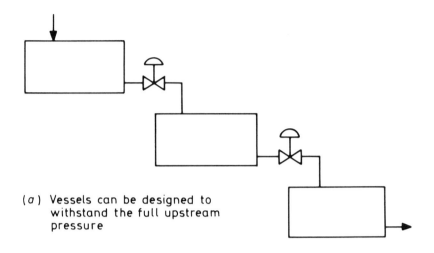

(*a*) Vessels can be designed to
withstand the full upstream
pressure

(*b*) Vessel can be designed to withstand pump closed-head
delivery pressure

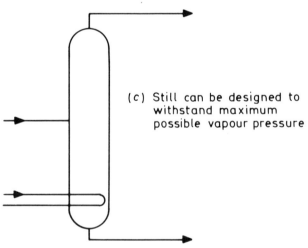

(*c*) Still can be designed to
withstand maximum
possible vapour pressure

FIGURE 1.1 Relief valves (but not fire relief valves) can be avoided by
using stronger vessels.

Figure 1.1(*a*) shows a series of vessels with a pressure drop between them. Relief valves are needed on the later vessels in case the letdown valves fail open and subject the vessels to the full upstream pressure. If all the vessels are made strong enough to withstand this pressure then relief valves are unnecessary.

Similarly, in Figure 1.1(*b*), if the vessel can withstand the pump delivery pressure then a relief valve is not needed.

In all these cases, if the equipment contains flammable materials then a fire relief valve will be needed.

References

1. Kletz, T.A. 1974. *Chem Processing* September, p. 7.
2. Kletz, T.A., and Lawley, H.G. 1975. *Chem Engineering* 12 May, p. 81.
3. Kletz, T.A. 1986. *Hazop and Hazan—Notes on the Identification and Assessment of Hazards,* 2nd ed., pp. 43, 44, 46. Rugby, England: Institution of Chemical Engineers.
4. Naughton, D.A. 1968. *Loss Prevention* 2:54.
5. Stewart, R.M. 1971. *Major Loss Prevention in the Process Industries.* Institution of Chemical Engineers Symposium Series No. 34. p. 99. Rugby, England: Institution of Chemical Engineers.

MYTH 2

A relief valve, properly designed, sized and maintained, will prevent a vessel from bursting. Special protection against overtemperature is unnecessary or at least a luxury.

A relief valve will prevent a vessel bursting if it is at, or near, its design temperature but not if the vessel gets too hot. Operating staff sometimes (though less often now than in the past) fail to realize that if a vessel gets too hot it may burst at or below the design pressure of the relief valve and that the relief valve provides no protection. The protection provided against overtemperature, if any, is usually primitive compared with the protection against overpressure provided by a relief valve.

Vessels can get too hot in several ways, and these are considered separately.

2.1 Vessels Heated by Electricity

Vessels that are heated by steam or hot oil usually cannot get too hot because they are normally designed to withstand the temperature of the heating medium, which is, of course, the maximum temperature attainable. Electric heaters, on the other hand, will overheat if the flow through them stops or gets too low. They are not self-regulating in the same way as steam or oil heaters. Heat input continues whatever the temperature. Protection against high temperature is therefore as necessary as protection against high pressure and, in this case, can be provided just as easily.

Nevertheless, several electric heaters have been installed (or originally designed) without any high-temperature trip operated by the shell temperature. (They may have been fitted with high-temperature trips to protect the heating elements, operated by the temperature of the heating elements, but the set points were too high to protect the shell.) This shows that the lack of protection against high temperature is due not to technical difficulties but to a "blind spot," a failure to realize that high temperature is dangerous. Even when a high-temperature trip is installed, it is usually easy to disarm it (i.e., render it inoperable), whereas disarming a relief valve is usually not possible.

2.2 Internally Insulated Vessels

Internally insulated vessels, such as reactors, are similar. The internal temperature is often higher than the shell will stand. Any deterioration of the insulation may overheat the vessel. Yet how often do we provide a high-temperature alarm or trip on the vessel wall?

2.3 Furnace Tubes

Vessels are usually tested to 1.5 times their design pressure and will usually withstand several times the design pressure before they burst. Often the vessel does not burst, but a flanged joint stretches and relieves the pressure. In contrast, furnace tubes will often withstand only a 5% or 10% increase in their absolute operating temperature without bursting. Nevertheless, the precautions taken to prevent bursting are usually primitive. Indirect protection may be obtained by a low-flow trip but these are not always fitted, are often easy to disarm and provide no protection against flame impingement. Direct protection can be obtained by tube skin temperature measurements, but these are usually considered insufficiently reliable to operate trips and are usually used only as warnings to the operator. Often we do not know what is the hottest part of the tube and so cannot fit the tube skin thermocouple in the right place.

It is admittedly difficult to protect furnace tubes against overtemperature with anything like the degree of reliability that we take for granted on overpressure protection. Our ignorance in this field is the result of the relative lack of effort expended in the past. If we had put so much effort into devising ways of protecting furnace tubes against excessive temperature as we have put into relief valve design, then the problems might have been solved. As it is, we failed to recognize that excessive temperature presents a hazard in many ways more serious than excessive pressure.

2.4   Vessels Exposed to Fire

In 1966 at Feyzin in France a leak of propane from a storage sphere caught fire. The Fire Brigade were advised to use the available water to cool neighboring spheres to prevent the spread of fire. It was assumed that the sphere on fire could be left to itself, because the relief valve would take care of it. As a result, after 1½ hours the sphere burst, killing 18 people and injuring many more.[1-6] The refinery staff had failed to appreciate that if a vessel gets too hot it will burst at or below the relief valve set pressure and that the relief valve will not prevent this from occurring. Initially the vessel was filled and the boiling liquid removed the heat. As the level fell, the unwetted upper portion of the vessel became heated.

Following this incident it was suggested that the incident had occurred because the relief valve was too small. It was, in fact, somewhat smaller than some companies would have used but this was not the cause of the incident.

Vessels that are exposed to a fire can be protected by:

1. Water cooling
2. Fire-proofing the vessel
3. Reducing the pressure
4. Sloping the ground

Water cooling has been the main line of defence adopted in the past, and the fire services are now usually fully aware of the need to apply cooling water as soon as possible.

Vessels have burst in as little as 10 minutes, so water must be applied quickly. Fixed equipment is therefore often necessary—either monitors that direct a large stream of water on the vessel or sprays that deliver a smaller quantity of water (10 $l/m^2/min$ or 0.2 $gal/ft^2/min$) over the whole surface.

Fireproof insulation is available as an immediate barrier to heat input, and does not have to be commissioned like water.

Sloping the ground ensures that any liquid that does not burn immediately runs off to one side. It must not be allowed to run onto the next plant. A collection pit may be necessary.

In addition it should be possible to lower the pressure in vessels exposed to fire. Often this can be done through existing process lines. Sometimes a relief valve bypass can be fitted and operated remotely. It is possible to obtain a combined relief valve and motor valve that can be lifted by remote operation but that also functions as a normal relief valve. Vapor depressuring should be provided on all vessels operating at a gauge pressure above 2 bar (30 psi), and the depressuring valve should be sized so that the pressure falls to 25% of design in 10 minutes, or 30 minutes if the vessel is insulated.

These methods of protection are illustrated in Figure 2.1, and more details can be found in Reference 7.

Although all four precautions are recommended, there is some trade-off between them. Thus if insulation is fitted the water rate can be reduced by two-thirds and the water can be poured over the vessel—it does not have to be sprayed at every part of the surface. Also, the depressuring valve (and relief valve) can be smaller.

As mentioned above, it is now more widely realized than in the past that relief valves will not protect a vessel that gets

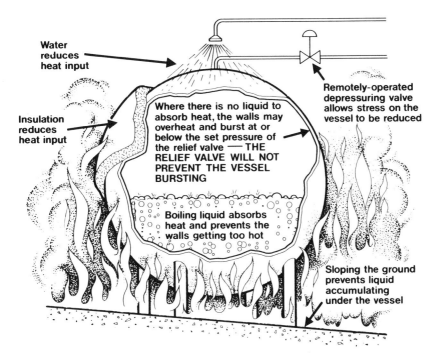

FIGURE 2.1 How to protect pressure vessels from fire.

too hot. I have discussed the Feyzin fire (or BLEVE—*boiling liquid expanding vapor explosion*—as it is often called) on many occasions with groups of operating and design engineers. In the early 1970s the groups would argue that because the vessel burst there must have been something wrong with the relief valve. Nowadays they usually realize quickly that the metal got too hot.

The British Standard for the design of unfired pressure vessels, BS 5500, first issued in 1976, requires vessels to be protected against overtemperature as well as overpressure (see quotation in Myth 1). Its predecessors, BS 1500 and BS 1515, did not include the reference to temperature.

2.5 Vessels that Get Too Cold

Relief valves will not protect vessels that get too cold. They may become brittle and fail, at or below design pressure, as the result of inherent stresses or external shocks. The leak and explosion at Beek in the Netherlands in 1975 occurred in this way.[8] Protection by trips may be necessary.

Some grades of steel become brittle at temperatures below
5°C, and vessels have failed while being pressure tested with
water below 5°C (see Myth 44).

References

1. *The Engineer,* 25 March 1966, p. 475.
2. *Paris Match,* No 875, 15 Jan 1966.
3. *Fire,* Special Supplement, Feb 1966.
4. *Petroleum Times,* 21 Jan 1966 p. 132.
5. Lagadec, P. 1980. *Major Technological Risk.* p. 176. Oxford,
   England: Pergamon Press.
6. Kletz, T.A. 1988. *What Went Wrong?—Case Histories of Process
   Plant Disasters*, 2nd ed., Section 8.1. Houston: Gulf Publishing Co.
7. Kletz, T.A. 1977. *Hydrocarbon Processing,* 56(8):98. (Reprinted
   in *Fire Protection Manual for Hydrocarbon Processing Plants.*
   Vol. 2, p. 291, ed. G.H. Vervalin. Houston: Gulf Publishing Co.)
8. van Eijnatten, A.L.M. 1977. *Chem Engineering Prog* 73(9):69.

MYTH 3

If a vessel is exposed to fire, it should be emptied as quickly as possible.

This is true if a leak from the vessel is feeding the fire—
draining the vessel to a safe place will remove the source of
fuel and extinguish the fire.

However, if a vessel is being heated by a fire and the fuel
is not coming from the vessel, it is usually safer to leave the
inventory in the vessel. The boiling liquid will remove the
heat. If the vessel is emptied, the metal will soon overheat,
the vessel will be damaged, and, if it is still under pressure, it
will burst violently.[1]

A domestic kettle is safe on a gas stove so long as it contains water. If it boils dry it is quickly damaged.

If a large number of joints are exposed to the fire—parti-
culary certain types of high-pressure joints such as lens ring
joints—it may be safer to empty the vessel. It is, of course,
necessary to have somewhere to put the contents, and some
high-pressure plants are provided with blowdown tanks for
this purpose.

It is also desirable to empty vessels containing materials that decompose explosively with rise in temperature.

Reference

1. Klaassen P. 1971. *Major Loss Prevention in the Process Industries*. Institution of Chemical Engineers, Symposium Series No. 34. p. 111. Rugby, England: Institution of Chemical Engineers.

MYTH 4

Using a vessel designed for a higher pressure is safer than using one designed for the operating pressure.

At first sight it seems obvious that a stronger vessel must be safer. However, although it is less likely to fail, if it does fail the damage is greater.

Suppose a vessel is needed for the storage of a liquid at or near atmospheric pressure. A low-pressure tank would normally be provided, but sometimes a redundant vessel is available and pressure is used.

Suppose the vessel is subjected to an excessive pressure because an internal coil bursts, gas is blown into the vessel, the vessel is overfilled, or an internal explosion occurs. The relief valve *should* be designed to guard against the first three but may not be. The low-pressure tank will probably have been designed for a pressure of 8 inches water gauge (4 kPa), and the roof will probably lift off at a pressure of 24 inches water gauge (12 kPa) but the contents will not spill. The pressure vessel *may* withstand the pressure, but if it does not, bits and the contents may fly in all directions.

Suppose the vessel is exposed to fire. The metal is softened and loses its strength (see Myth 2). The low-pressure tank bursts at a low pressure with little consequential damage. The pressure vessel, if not adequately vented, will burst at a higher pressure, again scattering the contents and bits of the vessel in all directions.

When vessels designed for a higher pressure than is necessary are used, their relief valves should be set just above operating pressure; they should not be set at the vessel's design pressure.

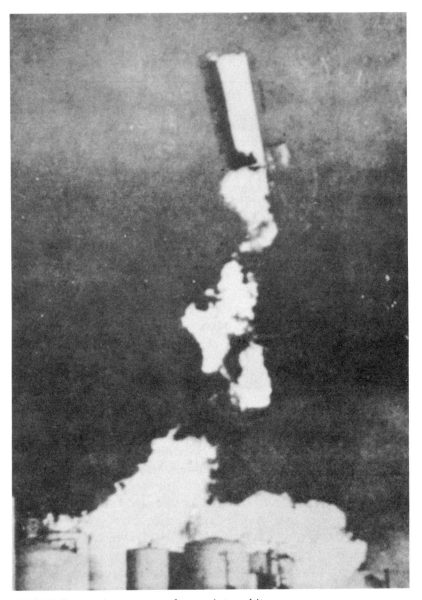

FIGURE 4.1   A storage tank goes into orbit.

The methyl isocyanate at Bhopal was stored in a pressure vessel rather than a low-pressure tank so that it could be transferred by nitrogen pressure. When a runaway reaction occurred the tank was distorted but did not burst. If it had been designed for a lower pressure it would have burst and the vapor would have been discharged nearer the ground. It would not have spread so far and fewer people would have been killed.[1]

An Incident

A tank truck hit a pipe leading to a storage tank and the pipe broke off inside the bund. The truck's engine ignited the spillage and started a bund fire that destroyed or damaged 21 tanks.

A 100-m³ vertical pressure vessel designed for a gauge pressure of 0.3 bar (5 psi), but used as a storage tank and vented into the atmosphere, contained 'Methyl Ethoxol' (the monomethyl ether of ethylene glycol). Its flash point is 40°C so it is not flammable at ordinary temperature, but the fire heated the liquid to its flash point and then ignited the vapor coming out of the vent. The fire flashed back into the vessel and an explosion occurred. The vessel came apart at the bottom seam and most of it took off like a rocket, with flames coming out of the base[2,3] (Figure 4.1).

Another Incident

When Guy Fawkes attempted to blow up the Houses of Parliament he laid stones and crowbars on top of the barrels of gunpowder. The weight, he knew, would make the explosion bigger.[4]

References

1. Kletz, T.A. 1988. *Learning from Accidents in Industry.* Chap. 10. Kent, England: Butterworths.
2. *Loss Prevention,* Vol 7, 1973, p. 119.
3. *Case Histories of Accidents in the Chemical Industry,* Vol 4, Item 1887, 1975. Washington, DC: Manufacturing Chemists Association.
4. Bowen, C.D. 1957. *The Lion and the Throne,* p. 208. London: Hamilton. p. 208.

MYTH 5

It is bad practice (and illegal) to fit a block valve below a relief valve but operators must be free to disarm trips that protect vessels from the effects of high or low temperatures, high or low levels, high concentrations of dangerous materials, and so on.

To deal with the legal side first, in the United Kingdom, if a vessel must be provided with a relief valve (see Myth 1) then a block (isolation) valve must not be fitted below it, unless two or more relief valves are provided and the block valves are interlocked. (The law is not very clear but is usually assumed to mean this.)

Many companies extend this legal requirement to all relief valves. Operators, they consider, make mistakes and may leave a block valve shut after a relief valve has been changed. This is a perfectly reasonable view. The same companies, however, may have an almost casual attitude to trips and alarms that protect equipment from the effects of high or low temperature, high or low levels, dangerous concentration of corrosive or explosive materials, and so on. Such trips may be easy to disarm, may be disarmed by operators on their own initiative, and may be left disarmed after trip testing, and there may be no system for indicating whether or not they are disarmed.

Why is it that disarming a relief valve fills us with horror but disarming a trip, which may protect against a more dangerous condition than overpressure, is often routine? There is no logical technical reason for the difference; we have unthinkingly absorbed the folklore of the industry.

If we accept that there is no logic in the present situation, what should we do? Should we treat trips like relief valves or relief valves like trips?

We should use the methods of hazard analysis[1,2] to estimate the probability and extent of a dangerous incident; that is, we should estimate

1. The probability that the relief valve will be called on to lift while it is isolated or the probability that the trip will be called on to operate while it is disarmed.
2. The severity of the consequences.

We should then compare the results of (1) and (2) with a target or criterion.

The results of such an analysis will show that we can safely treat some relief valves with less sanctity than in the past and fit block valves below them (see Section 5.3 for an example) and that many companies should treat trips with greater respect than in the past.

5.1   Occasions When Relief Valves Can Safely Be Fitted with Block Valves

1. When the maximum pressure that can be developed with the relief valve isolated will not take the equipment above its test pressure (or, more precisely, when the worst combination of pressure and temperature will produce stresses in the equipment less than those produced by the pressure test).
2. When the relief valve is protecting a long pipeline containing a liquid below its flash point (see Section 5.3) and one of the line isolation valves can be locked open.
3. When it is possible to lock open an alternative vent before isolating the relief valve.
4. When equipment isolation valves can be locked open so that the equipment is protected by another relief valve of adequate size.
5. When the demand rate on the relief valve is reduced to less than 0.1/year (once in 10 years) by an instrumented protective system that is regularly tested and maintained.

5.2   The Control of Trip Disarming

It may be necessary to disarm trips (and alarms) from time to time or alter their set points, but this should be done in a controlled manner. Operators should not be allowed to disarm or alter trips (or alarms) at will. The essential features of a control system are:

1. Disarming or alteration of trip or alarm settings must be approved in writing by an authorized person, normally a supervisor (plant manager in the United Kingdom) or senior foreman. Artificers should not alter trip or alarm settings without written authorization.
2. If a trip is disarmed, this should be signaled in some way, for example, by a light on the panel.
3. Trips and alarms should be tested regularly, say, every month. This is done primarily to detect failures but will also show if the trip (or alarm) has been disarmed or its setting altered.
4. If trips have to be disarmed for start-up (e.g., a low-flow trip) then they should rearm themselves automatically after a short period of time (several minutes to an hour depending on circumstances).

5.3 An Example of the Application of Hazard Analysis: the Isolation of Relief Valves

Relief valves on many interplant pipelines are difficult to release for testing. Should we therefore fit block valves beneath them, or is this too risky? Instead, should we go to the expense of twinned valves and interlocked block valves? As a result of the following calculation it was decided to install block valves below relief valves on pipelines handling liquids *below* their flash points.

The relief valves are installed to prevent the pipelines from being overpressured if they are isolated full of liquid and the temperature then rises. If a relief valve is left isolated in error then the pipeline will leak at a pair of flanges.

Assume:

1. There are 100 line relief valves in the area covered by a single operator (or four shift operators).
2. Each relief valve is isolated for inspection once every 2 years.

3. Once in every 100 occasions the procedure breaks down, one relief valve is left isolated, and this is not detected until the next inspection (i.e., on average, at any time one relief valve is isolated).
4. The 'demand rate' on each relief valve is once/year.
5. The operator is always near enough to one pipeline to be exposed to spray from a flange (i.e., for 1% of his time he is near the pipeline with the isolated relief valve).
6. The chance of the operator being hit by the spray if he is present is 1 in 10.
7. The chance of the operator being injured sufficiently seriously to result in a lost-time accident (LTA) is 1 in 5.

These assumptions are all based on the judgment of experienced people and they all err on the safe side.

If we accept these assumptions then it follows that:

1. One pipeline will be overpressured and will leak at a flange once in 2 years. (The period between tests is 2 years and there are two demands in this period, but after the first demand the fact that the relief valve is isolated will be discovered).
2. An operator will be exposed once in 200 years.
3. An operator will be hit by spray once in 2000 years.
4. An operator will be injured once in 10,000 years or once in 8.75×10^7 hours, equivalent to an LTA frequency rate of 0.001 (in 10^5 hours).

Most companies consider an LTA rate of 0.5 as good and 0.1 as outstandingly good.

A hazard that increases the LTA rate by 0.001 is so small that we should not allocate resources to reduce it further when there are so many much bigger risks to be dealt with first. Relief valves on pipelines carrying oils below their flashpoints may therefore be fitted with block valves.

Two points regarding this analysis should be noted:

1. Most examples of hazard analysis deal with fatal accidents. The method can also be applied to LTAs.
2. We have assumed that an administrative procedure will be set up to prevent block valves under relief valves from being left isolated but nevertheless we have assumed that sooner or later this procedure will break down and we have tried to estimate the frequency of breakdowns.

References

1. Kletz, T.A. 1986. *Hazop and Hazan—Notes on the Identification and Assessment of Hazards,* 2nd ed. Rugby, England: Institution of Chemical Engineers.
2. Lees, F.P. 1980. *Loss Prevention in the Process Industries.* Chap. 9. Kent, England: Butterworths.

MYTH 6

Designers can assume that operators will do what they are asked to do, provided they are properly trained and instructed and the task is within their mental and physical powers.

This myth is believed in more by designers than by those who operate plants.

Reports from all industries (and road accident reports) show that more than half the accidents that occur, sometimes 80% or 90%, are said to be due to human failing. If only people would take more care, or would follow instructions, then we would have fewer accidents.

True, but if people have not taken care or followed instructions in the past, we cannot assume that they will change their ways in the future. Engineers should accept men as they find them and design accordingly, and leave the changing of human nature to teachers and others more qualified to do so (who, judging by the results achieved in the past few thousand years, are not very successful in doing so).

A Story

A man went into a tailor's shop for a ready-made suit. He tried on most of the stock without finding one that fitted him. Finally, in exasperation, the tailor said, "I'm sorry, sir, I can't fit you. You're the wrong shape."

Reasons for Human Error

There are many reasons why people fail to do what we ask but, in the individual context, three predominate:

1. A moment's aberration
2. Poor training
3. Lack of supervision

6.1 A Moment's Aberration

Well-trained, well-motivated men, physically and mentally capable of doing what we ask, make occasional mistakes. They know what they should do, and want to do it, but occasionally forget to do it. For example, they forget to open a valve or open the wrong valve. Exhortation, punishment, or further training will have no effect. We must either accept an occasional mistake or change the work situation (that is, the plant or method of operation) so that errors are less likely. We should try to remove or reduce opportunities for error.

Note that errors occur not in spite of the fact that men are well-trained but because they are well trained. Routine activities are delegated to the lower level of the brain and are not continuously monitored by the conscious mind. We would never get through the day if everything required our full attention. When the normal flow or pattern of actions is interrupted, errors may occur.

For example, during the start-up of a unit an operator forgot to open a valve and as a result an explosion occurred. The unit and a similar one had between them been started up successfully 6000 times before the explosion occurred.[1] Nevertheless, one explosion in 6000 is too many. An interlock should have been installed so that the operator could not proceed with the start-up until the valve was open.

In some compressors it is possible to interchange suction and delivery valves. Damage and leaks have resulted when this has been done. Valves should be designed so that they cannot be interchanged.

Designers should not, of course, assume that operators will always make mistakes. They should try to estimate the probability of a mistake, and then decide, taking the consequences into account, whether to accept the occasional error or modify the work situation. Men are actually very reliable, but there are hundreds, perhaps thousands, of opportunities for error in the course of a day's work, so we should not be surprised that some errors occur.

Another opportunity open to designers is to allow the operator to recognize that he has made an error and correct it. For example, in entering instructions into a computer an operator sometimes has to check that the instructions are correct and then press the 'Enter' button a second time. Unfortunately operators may soon get into the habit of pressing the 'Enter' button twice without checking.

Slips are increased by stress and distraction and sometimes these can be reduced.

6.2   Poor Training

Many incidents have occurred because operators were not able to diagnose faults. Three Mile Island is the best known example.[2] In other cases operators have not understood how equipment works or did not understand what they were required to do. On a number of occasions operators have written down on record sheets temperatures or pressures that indicated an approaching hazard, but did nothing about them. Ultimately a dangerous incident occurred.

Designers should bear in mind that the plants they design will probably be operated by people who are just as prone to these errors as people have been in the past.

6.3   Lack of Supervision

Men carrying out routine tasks may become careless and take short cuts. Managers and supervisors cannot stand over them but should check, from time to time, that procedures are being followed. Unfortunately this does not always happen.

For example, as the result of corrosion on a plant in which water was electrolysed, some of the hydrogen entered the oxygen stream and an explosion occurred. The hydrogen and oxygen were supposed to be analyzed every hour for purity. After the explosion it was found that when plant conditions changed the oxygen analyses on the record sheet changed immediately, although it would take an hour for such a change to occur on the plant. Clearly, the analyses were not being carried out and the managers had not noticed this, and had failed to impress on the operators the importance of regular analyses and the serious consequences that could follow if the oxygen purity fell.[3]

Designers should not assume that the plants they design will be operated by supermen who never take short cuts or break the rules. They should assume that people will behave as they have done in the past. Their designs need not be foolproof, but they should be able to withstand without serious failure the sort of mistakes that experience shows will occur. To take a very simple example, any equipment installed on the plant is liable to be stood on. It is no use telling people not to stand on it. It should be made strong enough to stand on, or shaped or located so that it cannot be stood on (Figure 6.1).

For other examples see Myth 30.

6.4 Examples of Human Error from Other Industries

1. In the early days of anesthetics chloroform was often mixed with air and piped to a facemask using the apparatus shown in Figure 6.2, introduced in 1867. If the two pipes were interchanged liquid chloroform was supplied to the patient. This happened on a number of occasions, with fatal results.

    Redesigning the apparatus so that the two pipes could not be interchanged was easy but persuading doctors to use the new designs was another matter. They did not believe that a professional man could make such a simple error. As late as 1928 deaths from

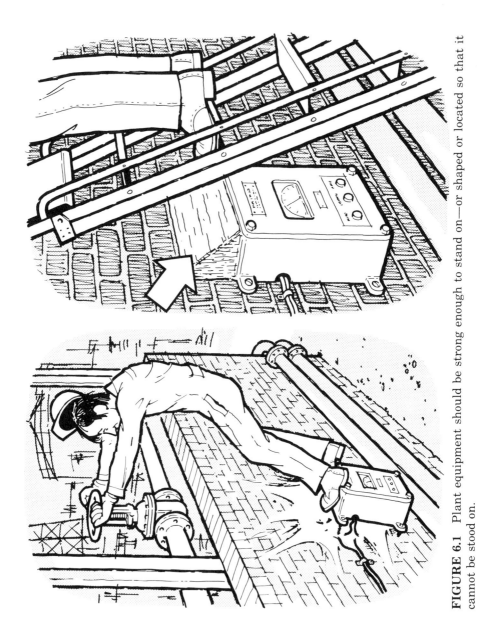

FIGURE 6.1 Plant equipment should be strong enough to stand on—or shaped or located so that it cannot be stood on.

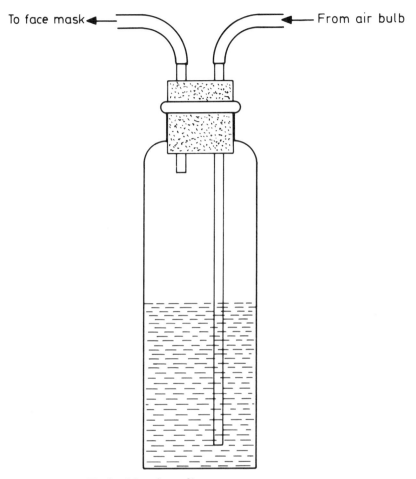

FIGURE 6.2   Early chloroform dispenser.

the use of the simple apparatus were still being reported.

    Many deaths could have been avoided if anesthe-
tists had used an apparatus that had been redesigned
so that errors could not occur.[4]

2. After a railway bridge had been constructed on the
    Georgetown Loop line in the Rocky Mountains in
    1883, it was found that the two supporting towers had
    been transposed. The north tower had been erected on
    the south bank and *vice versa,* so that the entire struc-
    ture had been built backward.

The bridge (of rivetted construction) had to be dismantled and rebuilt.[5]

For other examples of human error see Kletz (1985).

References

1. Vervalin, C.H., ed. 1985. *Fire Protection Manual for Hydrocarbon Processing Plants*. 3rd ed. p. 95. Houston: Gulf Publishing Co.
2. Kletz, T.A. 1988. *Learning from Accidents in Industry*. Chap. 11. Kent, England: Butterworths.
3. *The Explosion at Laporte Industries Ltd on 5 April 1975*. 1976. London: Her Majesty's Stationery Office.
4. Davies, W.S. 1982. *Essays on the First Hundred Years of Anaesthesia*. pp. 3–5. London: Longmans.
5. Morgan, G. 1976. *Rails around the Loop*. p. 15. Fort Collins, Col: Centennial Publications.

Further Reading

Kletz, T.A. 1976. *J Occup Accidents* 1(1):95.

Kletz, T.A. 1980. *Proceedings of the Third International Symposium on Loss Prevention and Safety Promotion in the Process Industries*. p. 2/205. Basle: Swiss Society of Chemical Industries.

Kletz, T.A. 1985. *An Engineer's View of Human Error*. Rugby, England: Institution of Chemical Engineers.

Kletz, T.A. 1988. *What Went Wrong?—Case Histories of Process Plant Disasters*, 2nd ed. Chap. 3. Houston: Gulf Publishing Co.

MYTH 7

Accidents are due to human failing, so we should eliminate the human element when we can.

This myth is the opposite to Myth 6. Both assume that accidents are due to human failing, but Myth 6 assumes that people can be persuaded to stop making mistakes, whereas

Myth 7 accepts that people always will make mistakes, and proposes we try to remove our dependence on them.

To deal first with the first part of the myth, "Accidents are due to human failing," as we showed in the discussion of Myth 6 this is true of many accidents, but it is not very helpful to say so. We cannot stop humans from failing so we should look for other ways of preventing accidents. In any accident investigation we should look for those causes that we can do something about.

Can we therefore eliminate the human element? Suppose that when an alarm sounds an operator has to go outside, select the right valve out of many, and close it within 10 minutes. He may fail to do so, if he is working in a busy control room, on one occasion out of 10 or, in a quiet control room, on one occasion out of 100.

We can eliminate the operator from the loop by arranging for the valve to be closed automatically when the alarm condition is reached. If we do this, we remove our dependence on the operator but we are now dependent on the men who design, install, test, and maintain the automatic equipment. They also make mistakes. We have not removed our dependence on men; we have merely transferred it from one man to another (or others). It may be right to do so, because the men who design, install, test, and maintain the equipment probably work under less stress and distraction than the operator, and, if so, their error rate will be lower. But do not let us kid ourselves that we have eliminated our dependence on the human factor.

MYTH 8

Trips are unreliable: the more trips we install, the more spurious trips we get, so on the whole it is better to rely on operators.

This myth is heard in America more often than in Europe. American companies on the whole seem less willing than European ones to rely on trips, and more willing to rely on operators. Unfortunately, as shown in the discussion of Myths 6 and 7, the reliability of operators may not be good enough.

The reliability of a trip can be increased to any desired level by increasing the test frequency or duplicating components. (But only up to a point. If we increase the test frequency too much, the time that the trip is out of action during testing becomes important. If we duplicate too many components, common mode failures become important.[1]) With this qualification, if we specify the reliability required, the instrument engineer can design a trip (or group of trips) with this reliability.

Duplicating a trip does increase the spurious trip rate, but this can be compensated for by installing a voting system. For example a duplicated trip might typically give a spurious trip every 0.75 year. A two-out-of-three voting system might have a spurious trip once in 100 years. The cost of the voting system can be justified by the savings that result from the avoidance of spurious trips.[1]

Note: In a two-out-of-three voting system there are three measuring instruments. Two of them have to indicate a hazard before the trip operates. Spurious trips—the trip operating when it should not because of a fault in the measuring equipment—are therefore much fewer.

Reference

1. Kletz, T.A. 1986. *Hazop and Hazan—Notes on the Identification and Assessment of Hazards.* 2nd ed., pp. 61, 63. Rugby, England: Institution of Chemical Engineers.

MYTH 9

If a flammable material has a high flashpoint it is safe and will not explode.

This is true so long as the material is below its flashpoint but once gas oil, fuel oil, heat transfer oils, and other high boiling point liquids are heated above their flashpoints they become as dangerous as gasoline and must be treated with the same respect.

In addition, a material will explode at as much as 50°C below its flashpoint if it is in the form of a finely divided mist.

At low pressures gases or vapors can explode below their normal flashpoints. For example, a jet fuel of flashpoint 50°C at atmospheric pressure has a flashpoint of 20°C at a pressure of 1 psi (7 kPa).[1,2]

Contamination by a small amount of low-flashpoint liquid will reduce the flashpoint considerably. For example, 2% gasoline reduced the flashpoint of a solvent from 37°C to 15°C (see 3 below).

Many fires and explosions have occurred because these facts, particularly the first, were not known to those concerned. For example:

1. A mixture of phenols was separated in a batch vacuum distillation column. The materials present all had flashpoints above 80°C and were considered safe. When a batch was complete the vacuum was broken with air. After a number of batches the steam coil in the boiler was withdrawn for cleaning. On one occasion, as it was being withdrawn an explosion occurred and the coil was shot out of the boiler, fortunately missing the men on the job.

    The coil had been withdrawn many times before without incident but on this occasion the maintenance team had started work rather sooner than usual after the completion of a batch and the distillation column was still hot, above the flashpoint of the vapors. The source of ignition was sparks or heat produced by withdrawing the coil.

    In the factory where this occurred gasoline was also handled. No one would have dreamed of adding air to equipment containing gasoline, nor of opening up the equipment while gasoline was still present. But the phenols were considered safe. Those concerned did not realize that when phenols are above their flashpoints they are as dangerous as gasoline and should be treated with the same respect.

    After the explosion, the vacuum was broken with nitrogen and the phenols were removed by adding wa-

ter and bringing it to a boil before opening up the
column to remove the heating coil.

Note: Do *not* add water to a vessel that is above
100°C. (see Myth 37).

2. A few years later, in the same factory, repairs had to
be carried out to the roof of a tank containing phenol.
The staff wished to avoid, if possible, emptying and
cleaning the tank but, remembering the previous incident, they emptied the tank as far as possible, isolated
the steam coil and allowed the few inches of phenol in
the base of the tank to set solid. (The melting point of
phenol is 41°C.) They then allowed a welder to weld a
patch on the roof.

The welder's torch vaporized and ignited some
phenol that had sublimed onto the roof and a mild
explosion occurred, lifting the roof but not removing it

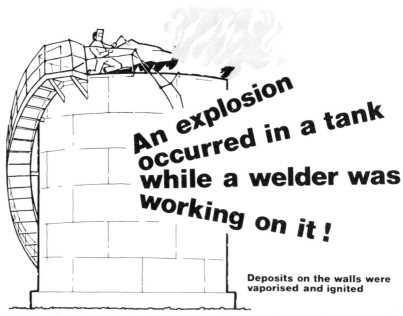

An explosion occurred in a tank while a welder was working on it !

Deposits on the walls were
vaporised and ignited

FIGURE 9.1 A tank that has contained heavy oils or solids can never be
made perfectly clean. The atmosphere inside must be made inert with nitrogen or inert foam before welding is allowed. Filling with water can reduce
the volume to be inerted.

completely. Fortunately the welder saw some fumes coming out of the vent and left the roof just before the explosion occurred.

3. A man was cleaning a moped engine using an apparatus in which a high-flashpoint (37°C) solvent was pumped from a drum to a cleaning brush, from which it fell into a tray and drained back into the drum. His clothes caught fire and, although he jumped into a pool of water, he later died from his burns.

It was found that a small amount of gasoline reduced the flashpoint of the solvent; 2% reduced it to 15°C. The source of ignition was never found with certainty, but may have been a spark from the circulating pump motor. It was not flameproof because it was intended for use with a high-flashpoint solvent.

Other explosions involving materials of high flashpoint are illustrated in Figures 9.1 and 13.3.

References

1. Penner, S. S., and Mullins, B.P. 1959. *Explosions, Detonations, Flammability and Ignition.* p. 119. Oxford, England: Pergamon Press.
2. Ministry of Aviation. 1962. *Report of the Working Party on Aviation Kerosene and Wide-Cut Gasoline.* London: Her Majesty's Stationery Office.

MYTH 10

Flammable mixtures are safe and will not catch fire or explode if everything possible has been done to remove known sources of ignition.

We are brought up to believe in the fire triangle: Air, fuel, and a source of ignition are necessary for a fire or explosion to occur; take one away and an explosion is impossible.

When flammable gases or vapors are handled on an industrial scale this view, though theoretically true, is misleading. If flammable gases or vapors are mixed with air in flammable concentrations, then experience shows that sources of ignition are likely to turn up. They are one of the few things in life we get free.

In many investigations of fires and explosions the source of ignition is never found. Sometimes the investigator attributes the ignition to static electricity but without demonstrating the precise way in which static electricity might have been responsible.

The amount of energy required to ignite a flammable mixture can be very small, as little as 0.2 mJ. This is the energy produced when a 1-cent coin falls 5 mm, though the energy has, of course, to be concentrated into a small time and space as in a spark or speck of hot metal. It is perhaps, therefore, not surprising that we cannot completely eliminate all sources of ignition.

As an alternative to the fire triangle, I suggest:

$$AIR + FUEL \rightarrow BANG$$
or
$$AIR + FUEL \rightarrow FIRE$$

What are these mysterious sources of ignition that seem to turn up? Sometimes it is genuinely static electricity. A steam or gas leak, if it contains liquid droplets or particles of dust, produces static electricity that can accumulate on an unearthed conductor, such as a piece of wire netting, a scaffold pole, or a tool. Discharges may perhaps occur from the cloud itself. In other cases ignition may be due to traces of pyrophoric material, to traces of catalyst on which reactions leading to local high temperatures may occur, to friction, or to the impact of steel on concrete (but not to the impact of steel on steel; see Myth 12).

The only safe rule is to assume that mixtures of flammable vapors in air in the explosive range will sooner or later catch fire or explode and should never be deliberately permitted, except under carefully defined circumstances where the risk is accepted. One such set of circumstances is in the vapor space of a fixed-roof storage tank containing a flammable conducting liquid such as acetone or methanol. Static electricity is not a serious risk, provided splash filling is not allowed, and experience shows that explosions are very rare. The same is not true of tanks containing flammable hydrocarbons with low flash points, and additional precautions such as nitrogen blanketing are necessary to reduce the risk of explosion to an acceptable level.

Here are just two examples of fires or explosions caused by unusual sources of ignition.

An explosion occurred in a fixed-roof storage tank. The liquid had a high conductivity, so static electricity could be ruled out as the source of ignition. The only source that could be found was frictional heating caused by a taut vibrating wire, which supported a swing arm, rubbing against a pulley that had seized on its bearing and was not free to move. Experiments have shown that steel wires subject to friction can produce glowing filaments of thin wire that cannot ignite methane but might ignite other gases.[1]

Reference 2 describes a fire in an open tank that occurred while a mechanic was tightening a screwed fitting that was leaking. The tank had been emptied but still contained flammable vapor. The tank was made from aluminum but the fitting—a valve—was made from steel. According to the report, friction between the steel and aluminum caused oxidation of aluminum to aluminum oxide, a reaction that is exothermic. Alternatively, a thermite reaction between aluminum and rusty iron might have occurred.

References

1. Kletz, T.A. 1988. *Learning from Accidents In Industry*. Chap. 6. Kent, England, Butterworths.
2. *Case Histories of Accidents in the Chemical Industry*. Vol. 4. Item 2184, 1975. Washington, DC. Manufacturing Chemists Association.

MYTH 11

The worst crime or mistake one can commit in a plant handling flammable liquids or gases is to introduce a source of ignition.

This is related to Myth 10. Reports on fires and explosions often show an excessive concern with the source of ignition. If it is discovered, it is often listed as the "cause." However, since sources of ignition usually turn up once a flammable mixture is formed, the real cause of the fire or explosion is the failure that allowed the flammable mixture to form, either by letting liquid or gas out of the plant or by letting air in. The only sure way of preventing fires and explosions is to keep the fuel inside the plant and the air out of it.

Which is the greater crime in a plant handling flammable liquids or gases: To bring in a box of matches or to bring in a bucket (Figure 11.1)? Most people would say that it is more dangerous to bring in the matches, but they are dangerous only if they are struck when a flammable mixture is present and in a well-run plant that is very rare. If a bucket is used, however, for collecting drips or samples or liquid for cleaning, a flammable mixture is always present above the liquid and may be ignited by a stray source of ignition (Figure 11.2).

I do not suggest that we should allow indiscriminate smoking, welding, and so forth in our plants. Obviously we must do what we can to remove known sources of ignition, so that those leaks that do occur are less likely to ignite, but this is our second line of defence. The first line is to prevent the formation of a flammable mixture. We should keep buckets out of our plants as conscientiously as we keep out matches.

For an account of a serious explosion that occurred because those concerned thought they had eliminated all sources of ignition and were therefore casual about leaks, see reference 1.

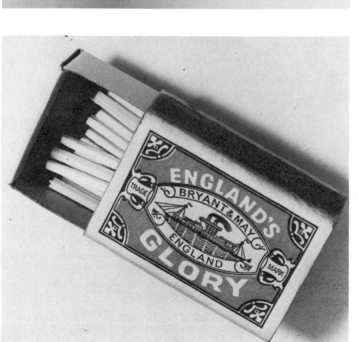

FIGURE 11.1 Which of these is more dangerous in a plant handling flammable liquids?

41

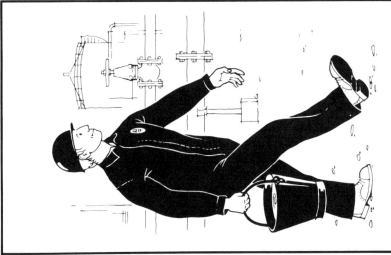

FIGURE 11.2 Buckets are more dangerous than matches.

42

Reference

1. Kletz, T.A. 1988. *Learning from Accidents in Industry.* Chap. 4. Kent, England: Butterworths.

MYTH 12

Nonsparking tools should be used in plants that handle flammable liquids or gases.

Nonsparking (better described as spark-resistant) tools seem to be regarded as a sort of magic charm to prevent explosions, though a series of reports over 30 years has shown that they have little value in this regard.

The American Petroleum Institute (API) has published a Safety Data Sheet[1] that summarizes these reports. It does not say when the tools were first introduced, but as far back as 1930 a number of engineers were asking if they were really necessary. In 1941 an API report showed that it was very unlikely that petroleum vapor could be ignited by the impact of steel on steel produced by hand, and that power operation is required to produce an incendive spark. Later investigations confirmed these conclusions. It may, be possible to ignite hydrogen, ethylene, acetylene, and carbon disulfide by the impact of steel on steel using hand tools, but we should not let anyone carry out a maintenance job in an explosive atmosphere of hydrogen, ethylene, or anything else.

If a leak is small, a man may be allowed to put his hands, protected by gloves, into a cloud of flammable vapor to harden up a leaking joint. Where hydrogen, ethylene, acetylene, and carbon disulfide are handled nonsparking hammers should be available for this purpose. It is probably better to use steel wrenches because they will harden up the joint more effectively.

If a flammable cloud extends more than a foot or so from the leak, no attempt should be made to harden up the leaking joint. Men should never be asked to put all or most of their bodies into a cloud of flammable vapor to carry out a repair job, because the vapor may ignite.

There is no harm in using nonsparking hammers for all purposes but it is an unnecessary expense. Care must be taken that small particles of grit do not get embedded in the hammers or they will be more dangerous than steel ones.

There is a need for nonsparking tools when explosive substances such as solid explosives are handled. The foregoing discussion applies to mixtures of flammable gas or vapor and air.

Reference

1. *Safety Data Sheet No PSD 2214.* 1973 (December). New York: American Petroleum Institute.

MYTH 13

If a combustible gas detector reads zero then there is no vapor present and it is safe to introduce a source of ignition.

Combustible gas detectors, both fixed and portable, are some of the most useful instruments we possess and have made a big contribution to safety. However, a knowledge of their limitations is essential if we are not to be misled by them. Unfortunately operators are often too willing to believe a zero reading, perhaps because that is the reading they would like to see.

Some of the causes of incorrect readings are:

1. *The instrument is out of order.* When these instruments fail they do not always fail safe. Portable instruments should therefore be tested every day or, better still, immediately before use every time they are used (Figure 13.1). A useful test material is 30% isopropanol in water, because it normally produces a low reading (57% of the lower flammable limit at 18°C) and will thus detect loss of sensitivity. It also prevents damage to the filament of the instrument by repeated exposure to rich mixtures.

Flammable gas detectors do not fail safe – so...

... test them before use – EVERY TIME

FIGURE 13.1

2. *The vapors may be absorbed by the sample tube.* For this reason it is better to use instruments in which the detecting element is placed at the point of test.
3. *The sample tube may be choked* as a result of the swelling caused by absorption of vapors or in other ways.
4. *The element may have been poisoned by exposure,* for example, to halogenated hydrocarbons or silicones. Poisoning by the former is temporary, by the latter permanent.
5. *The substance being detected may form a flammable mixture with air only when hot* and may cool down in the instrument. This is the most common cause of failure to detect explosive mixtures, and one explosion that occurred as a result will therefore be described in detail.

A furnace tripped out on flame failure as the result of a reduction in fuel oil pressure. The operator closed the two isolation valves and opened the bleed (Figure 13.2).

When the oil supply pressure had been restored the foreman tested the inside of the furnace with a combustible gas detector. He got no response and, therefore, inserted a lighted poker; a bang occurred, damaging the brickwork and slightly injuring the foreman.

When the burner went out, it took a few seconds for the solenoid valve to close, and during this time oil entered the furnace. In addition, the line between the valve and the burner may have drained into the furnace. The vapor from this oil (flashpoint 65°C) was too heavy to be detected by the combustible gas detector, because it condensed out in the sample tube. If a detector in which the detector head is placed at the point of test had been used the vapor might have condensed out on the sintered metal that surrounds the detector head.

When relighting a hot furnace burning fuel with a flash point above ambient temperature, we cannot rely on a combustible gas detector to detect a flammable mixture. We should therefore sweep out the furnace for a long enough period of time to be certain that any unburnt oil has evaporated.

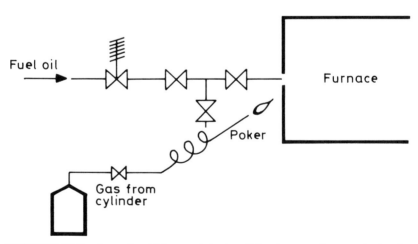

FIGURE 13.2   A combustible gas detector did not prevent an explosion in this furnace.

FIGURE 13.3 Heavy oils, which are not detected by gas detectors, can explode when they are heated above their flash points.

Operators should know the reason for purging so that they are less likely to reduce the purge time to avoid delay.

It is not a bad rule to say, "If a furnace burning fuel oil trips, have a cup of tea before relighting it." This will give most furnaces time to purge. If the delay is unacceptable then permanent pilot burners, supplied from a separate fuel supply, may be used.

To keep the purge time as short as possible, the solenoid valve should close quickly, it should be close to the burner, and the line in between should be sloped so that it does not drain into the furnace.

There is a need for a gas detector that can detect vapor that is explosive when hot but safe at atmospheric temperature, for example, a detector with a heated sample tube.

For many years the furnace on which the explosion occurred and other furnaces burning heavy oils had been tested with detectors that were incapable of detecting the vapor of the fuel. Often we disbelieve instruments; we believe them uncritically when they tell us the answer we want.

Another incident is illustrated in Figure 13.3.

MYTH 14

Gasoline engines produce sparks and must not be used in areas in which flammable gases and vapors are handled, but diesel engines are OK.

Although diesel engines do not use sparks they can ignite flammable mixtures of gas and vapor in air in other ways:

1. The flammable mixture can be sucked into the engine via the air inlet, and can ignite and flash back.
2. Sparks or flames can be produced by the exhaust, or the exhaust manifold or pipe can be hot enough to ignite the mixture.
3. The use of a decompression control can ignite a flammable mixture by exposing it to the hot gases in the cylinder.
4. Ancillary equipment may produce sparks.

In the incident described under Myth 4 a spillage was ignited by a diesel engine but the precise mechanism is not known.

In another incident, there was a spillage of 4 tons of hot cyclohexane. It occurred during a plant shutdown and a diesel-engined vehicle was operating in the area. The vapor was sucked into the air inlet and the engine started to race. The driver tried to stop it by isolating the fuel supply but without success because the fuel was going in with the air. Finally valve bounce and flashback occurred and ignited the cyclohexane. Four men were killed by the fire.[1]

Proprietory devices are available for protecting diesel engines so that they can operate safely in areas where leaks may occur. They operate by shutting off the air supply as well as the fuel supply, and can be operated manually or automatically when a leak occurs.[2] It is, of course, also necessary to ensure that flammable vapors cannot be ignited in other ways. A spark arrestor and flame trap should be fitted to the exhaust and it should be kept below the auto-ignition temperature of the materials used in the area; the decompression control, if fitted, should be disconnected so that it cannot be used, and electrical equipment should be protected.[3]

Note that if insulation is used to keep the temperature of the exhaust below the auto-ignition temperature of the materials handled, contact of liquid with the insulation (or even with dust that has settled on the exhaust pipe) may lower the auto-ignition temperature of the liquid.[4]

In June 1989 the press reported that a diesel railway engine had ignited a leak from a liquefied natural gas pipeline near Ulfa, Central Russia. Over 600 people were killed.

References

1. Kletz, T.A. 1988. *Learning from Accidents in Industry*. Chap. 5. Kent, England: Butterworths.
2. *Hazardous Cargo Bulletin*, Jan 1983, p. 22.
3. Oil Companies Materials Association. 1977. *Recommendations for the Protection of Diesel Engines Operating in Hazardous Areas*. Chichester, England, John Wiley & Sons.
4. Stanton, R.C. 1976. *Chem Engineer*, May, p. 391.

MYTH 15

If the pressure on a liquefied gas is reduced the amount of liquid remaining can be calculated by heat balance.

If a liquid is under pressure at a temperature above its atmospheric pressure boiling point and the pressure is reduced, then some of the liquid will flash and the rest will fall to its boiling point at the new pressure. It is easy to calculate the amount that will flash and the amount of liquid that remains. However, the flash is accompanied by the production of a great deal of spray, and experiments have shown that under the right conditions all the liquid may form a mixture of vapor and spray.[1,2] This phenomenon is well known to anyone who has removed the cap from the radiator of a car while the engine was hot. In this case the spray consists of large droplets that soon fall to the ground. In other cases the spray may be fine.

No one, so far as I know, understands in detail the factors controlling the amount or fineness of the spray. It probably depends on the liquid density, its surface tension, its degree of superheating and the geometry of the vessel and of the aperture causing the reduction in pressure.

Note that if the vapor is flammable a fine liquid spray is also flammable and remains flammable below the flash point (see Myth 9).

It is common practice to assume that the amount of spray produced is roughly equal to the amount of vapor produced (unless the vapor amounts to more than half the original liquid, in which case it is assumed that the rest of the liquid will form spray).

References

1. Reed, J.D. 1975. *Proceedings of the First International Symposium on Loss Prevention and Safety Promotion in the Process Industries,* p. 191. Amsterdam: Elsevier.
2. Osigo, C., et al. 1972. *Proceedings of the First Pacific Chemical Engineering Congress.* Part II, Paper 9–4. Tokyo.

MYTH 16

A pressure of 10 pounds is small and will not cause injury.

There is not a printer's misprint in the heading. I wrote it that way because that is how we usually speak. We say, "The pressure in the vessel is 10 pounds" because "10 pounds per square inch" is too much to say.

Unfortunately this leads to a belief that, because 10 pounds is not much, so a pressure of 10 pounds is not much. Once we analyze the myth we see it is wrong, but it is still hard to get the idea out of our heads. Whenever people have been injured or plant damaged by pressure, surprise is expressed—by technically qualified people as well as operators—that so little pressure could cause so much damage or injury.

For example, some years ago an operator opened the door, 3 feet 6 inches (1.1 m) diameter, of a steam filter before blowing off the pressure. The operator was crushed by the door against the frame of the filter and was killed instantly. During the investigation, surprise was expressed that such a small pressure (a gauge pressure of 30 psi or 2 bar) could cause the injuries and damage that occurred and a chemical explosion in the filter was suggested. In fact, simple calculation shows that the force acting on the door was 18 tons—and it is not surprising that when the holding bars were released it flew open with great violence.

In another incident a driver opened the manhole on top of a pressure tank truck wagon while there was air inside at a gauge pressure of 10 psi (0.3 bar). He was blown off the top. Surprise was expressed that the pressure was sufficient to do this.

See Myth 30 for more details of these two incidents.

On several occasions tank trucks have been emptied with the manhole and vents shut and have collapsed. Surprise has been expressed that the atmospheric pressure is sufficient. In fact, atmospheric pressure acting over the surface area of

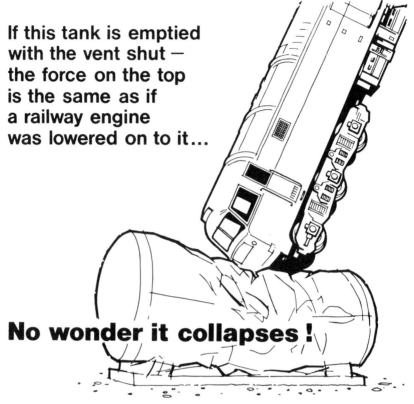

If this tank is emptied with the vent shut — the force on the top is the same as if a railway engine was lowered on to it...

No wonder it collapses !

FIGURE 16.1 The effect of emptying a tank with the vent shut.

even a small tank can amount to 200 tons. No one would expect the tank to survive if a railway engine (approximately the same weight) was lowered on to it (Figure 16.1).

The use of SI units, in which force and pressure are measured in entirely different units, may help to avoid confusion between them.

MYTH 17

Vessels are strong and can withstand any treatment they are likely to receive.

This is a myth believed by many plant operators and even by supervisors whose background is chemistry rather than engineering.

A pressure vessel originally designed for a gauge pressure of 0.3 bar (5 psi) was used for storing at atmospheric pressure, as liquid, a product with a melting point of 97°C. It was kept hot by a steam coil. The inlet line to the tank was being blown with compressed air at a gauge pressure of 5 bar (75 psi) to prove that it was clear, the usual procedure before filling the tank. The vent on the tank was choked and the end of the tank was blown off, killing two men who were working nearby.

The operators found it hard to believe that a 'blow of air' could burst a steel pressure vessel, and explosion experts had to be brought in to convince them that there had not been a chemical explosion. In fact, the air pressure was far greater than the bursting pressure of the vessel (a gauge pressure of about 1.3 bar or 20 psi). Steel looks stronger than air and, though we may be intellectually convinced that it is not, it is harder to take this for granted.

The vent on the vessel had choked on previous occasions and the operators had complained about the poor access, which made it difficult for them to see if it was clear and to rod it free if it was not. However, this choking was looked upon, by operators and supervisors, as an inconvenience rather than a hazard, and improvements carried no priority and were never made.[1]

Reference

1. Kletz, T.A. *Learning from Accidents in Industry*. Chap. 7. Kent, England: Butterworths.

MYTH 18

Blast walls provide the ultimate protection against explosions.

Despite all our precautions, explosions may occur inside or outside the plant equipment. Certain items of equipment in which the chance of an explosion occurring has been judged to be greater than normal (e.g., certain oxidation reactors) have been surrounded by blast walls to protect people and other equipment from missiles and blast.

The walls may give some people a feeling of security, but that is about all. To withstand the sort of pressures that might be developed, especially in a confined explosion, the walls would have to be so thick that they would cost as much as the plant. If an explosion occurred inside many so-called blast walls, the result would be that people would be hit by a stream of moving concrete instead of a stream of moving air. Even if the wall withstands the shock wave it may deflect pressure onto people sheltering behind it.

Instead of building blast walls we would do better to spend our money on reducing the probability that an explosion will occur. The high-integrity protective system described by Stewart[1] is better value for money than a blast wall because it will prevent explosions in oxidation reactors instead of providing doubtful protection against the consequences.

Equipment that is particularly liable to leak and fire is sometimes surrounded by fire walls, which prevent the fire reaching other equipment. These walls serve a useful purpose. They are often spoken of loosely as blast walls but they are really fire walls.

In small research plants, where the maximum energy release is equivalent to a few kilograms of TNT, construction of reliable blast walls is feasible and is the recommended solution, because a high-integrity protective system would be too expensive and often we do not have the knowledge needed to design one correctly. A method of designing blast walls for small units has been described by High.[2]

References

1. Stewart, R.M. 1971. *Major Loss Prevention in the Process Industries*. Institution of Chemical Engineers Symposium Series No 34, p. 99. Rugby, England: Institution of Chemical Engineers.
2. High, W.G. *Chemistry and Industry,* 3 June 1967, p. 899.

MYTH 19

Drums are safer than tanks for the storage of flammable liquids, because each one contains so little.

This myth seems to be believed by those responsible for writing codes and regulations as 100 tons of flammable liquid in drums can be stored nearer to a building than the same quantity in a tank.[1] Procedures have to be gone through and codes followed before a new tank is built, but stacks of drums are allowed to grow on spare ground. A few drums are put there 'temporarily' when stocks are high and before long they are a permanent part of the scene.

In fact, flammable liquid in drums is much more hazardous than the same quantity in a storage tank. It would be difficult to find a better way of burning liquid quickly than stacking it in thin-walled containers with spaces between them. In a fire in 1973, 90 tons of solvent in drums burned in 9 minutes[2,3] (Figure 19.1). In contrast, if a storage tank is exposed to fire, the roof may lift but the liquid is still confined in the tank and so is the fire (assuming that the tank is designed so that the roof/wall seam is weaker than the base/wall seam and will give way first).

Storage of aerosol containers is also more hazardous than is generally realized. In 1982 a large single-story warehouse (floor area 110,000 m²) was destroyed by a fire that started in a stack of cardboard boxes about 5 m high containing aerosol cans. Although the warehouse was protected by sprinklers, they were quite incapable of controlling the fire, and in addition rocketing aerosol cans spread the fire. The rocketing cans passed through water curtains that were intended to isolate sections of the warehouse.[4]

References

1. U.K. Health and Safety Executive. 1977 (January). *The Storage of Highly Flammable Liquids*. Guidance Note CS/2. London: Her Majesty's Stationery Office.

FIGURE 19.1 The scene after a fire in which 90 tons of solvent in drums burned in 9 minutes. Note that the drums were stacked close to buildings that were damaged or destroyed by the fire.

2. *Fire,* Dec 1973, p. 362.
3. *Fire Prevention,* May 1974, p. 40.
4. *Record* (published by Factory Mutual Insurance), Vol. 60, No. 3, Fall 1983, p. 3.

MYTH 20

Ton for ton, toxic gases produce more casualties than flammable gases or liquids

The concentrations of toxic vapor that can cause sudden death or injury are much lower than lower flammable limits. The spread of toxic vapors cannot be cut short by ignition. Hence we would expect sudden releases of toxic vapors to produce many more casualties than sudden releases of flammable vapors.

In practice, however, this is not the case. For the period 1970–75 the press, including the trade press, reported 34 fires or explosions in the oil and chemical industries (including transport) throughout the world that resulted in five or more fatalities. These amounted to about 600 fatalities in total. I know of only two comparable toxic incidents in the same period causing a total of 28 fatalities. They were: (1) an explosion in a refrigerated store that was caused by a leak of natural gas and that resulted in rupture of several ammonia tanks (it is not clear from the report whether the 10 fatalities were caused by the explosion or the ammonia[1]); and (2) a tank burst that killed 18 people.[2]

Simons[3] compared fatalities caused by the transport of flammable and toxic gases in the United States:

> During the period 1931 to 1961 37 persons (non-workers) were killed in LP-Gas flash fires and explosions from accidents involving tank trucks. This is an average of 1.23 fatalities per year . . . . For the years since 1961, an exact tally has not been made, but the annual average is believed to be in the range 1 to 2 fatalities per year.

In addition an unknown number of people were killed in accidents involving LP-gas tank cars.

In contrast, during the same period five people were killed in the United States as the result of accidents involving tank trucks and tank cars of chlorine. So far as is known, the transport of other toxic gases caused no fatalities in the United States in this period. Five deaths in 45 years is an average of 0.1 death/year.

The total quantity of flammable flashing liquid in stores, process plants, and transport containers probably exceeds the total quantity of toxic flashing liquid, but this is not sufficient to explain the difference. Marshall[4] estimated Mortality Indices: the average number of people killed by the explosion of a ton of hydrocarbon or the release of a ton of chlorine or ammonia. He finds that the historical record is:

Substance	Mortality Index
Chlorine	0.30
Ammonia	0.02
Liquefied flammable gases	0.60
Unstable substances	1.50

Why do sudden releases of toxic vapors kill so many fewer people than expected, despite the serious results that are theoretically possible?

The explanation may be that, while in theory a toxic vapor can spread a very long way and cause many casualties, weather conditions have to be exactly right and this rarely coincides with a leak. In addition, people can often escape; Simons et al.,[5] discussing chlorine, wrote: "People flee instinctively when confronted by the greenish, choking cloud. Flammable gas clouds do not provide such a clear warning of danger." Furthermore, if windows are closed a toxic gas cloud can pass over houses without causing casualties. On average, therefore, toxic vapors produce fewer casualties than expected.

Another reason why toxic vapors produce fewer casualties than estimated is that the data for the toxicity of chlorine used by many workers are almost certainly wrong by an order of magnitude.[6]

Taylor[7] discussed the reasons why poison gas was not used in World War II and wrote: "Most probably the explanation was the simple calculation that, weight for weight, high explosive was more effective than gas in killing people."

Unfortunately, at Bhopal, in 1984, weather conditions were exactly right and thousands of people were living in a shanty town close to the factory. The tumbledown dwellings provided no barrier to the entry of the poisonous gas (methyl isocyanate) and at least 2000 people were killed.[8]

References

1. *The Times,* 19 Sept 1973.
2. *Proceedings of Public Enquiry into Explosion of an Ammonia Storage Tank at Potchefstroom, South Africa on 13 July 1973.* Pretoria, South Africa.
3. Simons, J.A. 1975, *Risk Assessment Method for Volatile Toxic and Flammable Materials.* 4th International Symposium on Transport of Hazardous Cargoes by Sea and Inland Waterways, 26–30 October, Jacksonville, Florida.
4. Marshall, V.C. 1982. In *Hazardous Materials Spills Handbook,* eds. G.F. Bennett, F.S. Feates, and I. Wilder. New York: McGraw-Hill.
5. Simons, J.A., Erdman, R.C., and Naft, B.N. 1974. *Risk Assessment of Large Spills of Toxic Materials.* National Conference on Control of Hazardous Material Spills, San Francisco.
6. *Chlorine Toxicity Monograph.* 1987. Rugby, England: Institution of Chemical Engineers.
7. Taylor, A.J.P. 1965. *English History 1914–1945.* Oxford, England: Clarendon Press, p. 427; and Penguin Books, p. 524.
8. Kletz, T.A. 1988. Learning from Accidents in Industry, Chap. 10. Kent, England: Butterworths.

MYTH 21

Compressors and distillation columns are complex items of equipment and need our best operators, but furnaces do not.

In many companies the new operators are assigned to furnace operation. When they have proved their ability they may

If you let your furnace tubes run 60° C
hotter than design for 6 weeks, you
may halve the life of the furnace.

FIGURE 21.1   The effect of temperature on the life of a furnace tube.

be promoted to the control of the compressors or distillation columns. This implies that there is not much to furnace operation—anyone can do it.

This attitude has resulted in many expensive furnace tube failures—the result of operators (and sometimes supervisors) not fully understanding the way they work. In particular, they do not understand the behavior of the metal from which the furnace tubes are made.

Suppose the tubes are designed to operate at 500°C for 100,000 hours (11 years):

If they are operated at 506°C they will last 6 years.
If they are operated at 550°C they will last 3 months.
If they are operated at 635°C they will last 20 hours.

In each case failure will be by 'creep'—the tube will expand, slowly at first and then more quickly, and will finally burst.

If the tubes are operated at 550°C for 6 weeks they will use up half their creep life and will fail after 5 or 6 years instead of 11 years. No matter how gently they are treated, once they have been overheated they will never forget. Furnace tubes have a better memory than an elephant (Figure 21.1).

If a compressor or distillation column has a throughput greater than design then we can use the extra capacity. However, if we get more out of a furnace by turning up the burners we may pay for it later.

Other incidents involving furnaces have occurred because operators did not fully understand the principles underlying the lighting-up procedure and were tempted to take short cuts.[1]

A casual attitude to furnaces is not new. In the 19th century boiler explosions were frequent but engine breakdowns were much less common. This was due "in particular to the care taken in maintenance which even nowadays is in sharp contrast to the frequent neglect of the steam generator."[2]

References

1. *Furnace Fires and Explosions.* Hazard Workshop Module 5. Rugby, England. Institution of Chemical Engineers.
2. Eyers, J. 1967. *Maschinen Schaden* 40(1):3.

MYTH 22

If we want a piece of equipment repaired, all we need do is point it out to the man who is going to repair it.

Unfortunately, experience shows that the repairman goes to get his tools or finish another job and when he returns he cuts open the wrong pipeline, unbolts the wrong joint, or dismantles the wrong pump. Chalk marks are no better—they get washed off by rain or there are so many of them from past jobs that the repairman goes to the wrong one.

The recommended method for identifying equipment that is to be repaired is to attach a numbered tag to the equipment at the point of repair. If a pipeline has to be cut, for example,

Tag along with us

They make the job exact and neat...
Please return when job complete.

FIGURE 22.1 Equipment that is to be maintained should be identified by a numbered tag.

There were seven pumps in a row

A fitter was given a permit to do a
job on No. 7. He assumed No. 7
was the end one and dismantled it.
Hot oil came out.

The pumps were actually numbered:

Equipment which is given to
maintenance must be labelled.
If there is no permanent label then
a numbered tag must be tied on.

FIGURE 22.2 Permanent numbers should follow a logical sequence.

the tag should be attached at the point at which it is to be cut. The tag number should be put on the permit-to-work that is given to the repairman (Figure 22.1). If equipment has a permanent number painted on it or attached to it, this should be used instead of the tag. Permanent numbers should follow a logical sequence (Figure 22.2).

Further Reading

Kletz, T.A. *Hazards in chemical system maintenance: permits.* In *Safety and Accident Prevention in Chemical Operation.* Chap. 36, eds. H.H. Fawcett and W.S. Wood. New York: John Wiley & Sons.

Kletz, T.A. 1988. *What Went Wrong?—Case Histories of Process Plant Disasters*, 2nd ed. Chap. 1. Houston: Gulf Publishing Co.

MYTH 23

We should do all we can to remove hazards.

Twenty years ago this statement would have been accepted without hesitation. Safety was a black-and-white affair. If a hazard was recognized, certainly if an accident had occurred, then action had to be taken to prevent the accident happening again. Since resources are limited this often resulted in lavish spending to remove hazards that had been brought to our attention, by an accident or in other ways, accompanied by a reluctance to look too hard for other hazards in case we found more than we could deal with.

Now, to a large extent, the chemical industry has come to realize that we should search systematically for hazards, using techniques such as hazard and operability studies,[1,2] and then use a systematic technique, preferably numerical, such as hazard analysis,[1,3] to decide which hazards should be dealt with immediately and which can be left, at least for the time being.

The Canvey Island Reports[4,5] (quantitative estimates of risk to the public from the oil and chemical plants on an island in the Thames estuary, near London) were milestones in the adoption of these techniques. The methods, data, and criteria used in the Reports have been criticized but nevertheless they demonstrate an acceptance by government that one cannot do everything possible to prevent every conceivable accident and therefore we should try to quantify the risks and compare them with a target or criterion.

References

1. Kletz, T.A. 1986. *Hazop and Hazan—Notes on the Identification and Assessment of Hazards.* 2nd ed. Rugby, England: Institution of Chemical Engineers.
2. Lees, F.P. 1980. *Loss Prevention in the Process Industries.* Chap. 8. Kent, England: Butterworths.
3. Lees, F.P. 1980. *Loss Prevention in the Process Industries.* Chap. 9. Kent, England: Butterworths.
4. *Canvey, An Investigation of Potential Hazards from Operations in the Canvey Island/Thurrock Area.* 1978. London: Her Majesty's Stationery Office.
5. *Canvey—A Second Report.* 1981. London: Her Majesty's Stationery Office.

MYTH 24

We can remove hazards.

This is related to the last myth—if we say we should remove all hazards, we imply that we are able to do so. Unfortunately we cannot always do so, for several reasons:

1. Safety is often approached asymptotically. There is a small chance that a relief valve will fail and a vessel will be overpressured. This chance can be made even smaller by fitting two full-size relief valves, but the risk of overpressuring is still not zero because coincident failure is possible.[1]
2. The second reason that we will never remove all hazards is that we shall fail to foresee all of them. We do try to foresee what might go wrong by the use of tech-

niques such as hazard and operability studies,[1] but the people who use the techniques are not omniscient, and some hazards will be missed. In particular we may fail to foresee the consequences of modifications (see Myth 25).

3. The third reason we cannot remove all hazards is that men will make occasional mistakes and we cannot remove our dependence on men by installing automatic equipment. All we can do is to transfer our dependence from one man to another (see Myth 7).

Reference

1. Kletz T.A. 1986. *Hazop and Hazan—Notes on the Identification and Assessment of Hazards.* 2nd ed. Rugby, England: Institution of Chemical Engineers.

MYTH 25

The successful man is the one who gets things done quickly.

At one time the 'go-getter' who brushed difficulties aside was considered the successful man, but no longer. When plant or methods of operation are changed, those who rush repent at leisure. Changes to plants and methods of operation often have unforeseen side effects. The successful man is now the one who says "I know it's urgent. But we can afford to spend an hour or two looking critically and systematically at the proposed modification." This is quite a cultural change.

Of course, we want ideas for change as much as ever, but we also want to try to find out all their consequences before we go ahead.

The following are some examples of simple changes to plants or methods of working that had unforeseen and unwanted side effects:

1. When storage tanks are exposed to fire they are kept cool by pouring water over them, from fixed installa-

tions or mobile monitors. The water is poured or directed onto the roof and runs down the sides.

In 1968 a new U.K. Standard (BS 2654, Part 3) allowed tanks to be made with thinner walls than before, provided the walls were reinforced by wind girders. When the first tanks were constructed according to the new design it was found that the girders prevented the cooling water running down the sides of the tanks. Special deflection plates had to be fitted to redirect the water on the walls.

A better solution, if the consequences had been foreseen, would have been to put the girders *inside* the tanks (Figure 25.1).

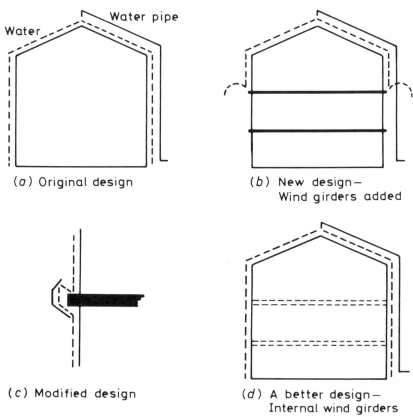

FIGURE 25.1   The unforeseen results of strengthening a tank with wind girders.

2. A vent line was arranged as shown in Figure 25.2(*a*). When repairs were made a straight-through cock (*b*) was not available, so a right-angle type was used [Figure 25.2(*b*)]. When the vent was used the reaction forces caused it to whip round, as shown in Figure 25.2(*c*). Fortunately, no one was hurt, though a similar movement of a vent pipe has caused a fatal accident.[1]

    If a right-angled cock had to be used it should have been installed as shown in Figure 25.2(*d*).

3. Large containers, about 60 m^3 in volume, had to be filled with powder. A large plastic bag was put into each container and inflated with nitrogen. The normal practice was to put the nitrogen hose inside the neck of the plastic bag, but one day, to try to save time, an operator tied the neck to the hose with string (Figure 25.3). Fifteen minutes after turning on the nitrogen there was a loud crack and the top of the metal container was found to have bowed out by about 3 inches.

    The plastic (0.007 inch thick) is not, of course, stronger than the steel (0.064 inch thick) but the plastic bag was bigger than the steel container so the steel gave way first.

    The operator should not have connected the hose to the plastic bag without first checking that the container and bag could withstand the full pressure of the nitrogen.

4. The most famous modification ever made in the chemical industry was the installation of a temporary pipe at the Nypro factory at Flixborough in 1974. It failed 2 months later, causing the release of about 50 tons of hot cyclohexane, which mixed with the air and exploded, killing 28 people and destroying the plant.[2–4]

    On the plant there were six reactors in series; each was slightly lower than the one before so that the liquid in the them flowed by gravity from No. 1 down to No. 6 through short, 28-inch diameter connecting pipes [see Figure 25.4(*a*)]. To allow for expansion each 28-inch pipe contained a bellows.

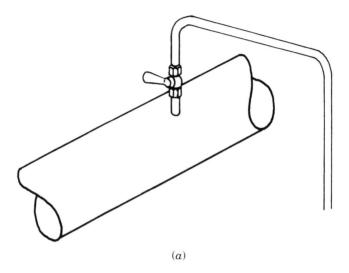

(a)

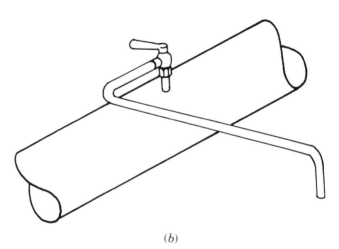

(b)

FIGURE 25.2 (a) Original arrangement of vent pipe. (b) When repairs were made a straight-through cock was not available, so a right-angle type was used. (c) When the vent was used the reaction force caused it to whip round. (d) If a right-angled cock had to be used, it should have been installed as shown here.

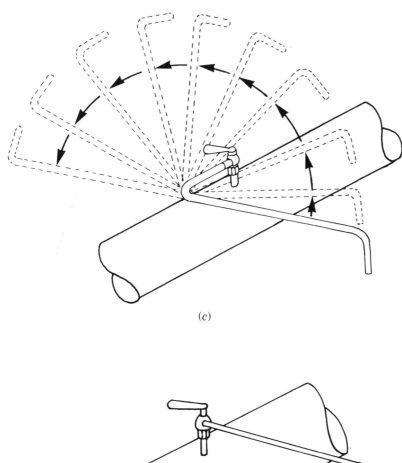

(c)

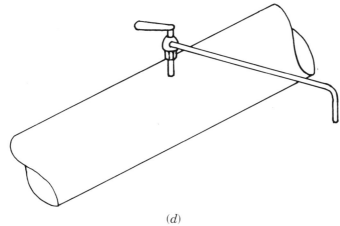

(d)

FIGURE 25.2 (continued)

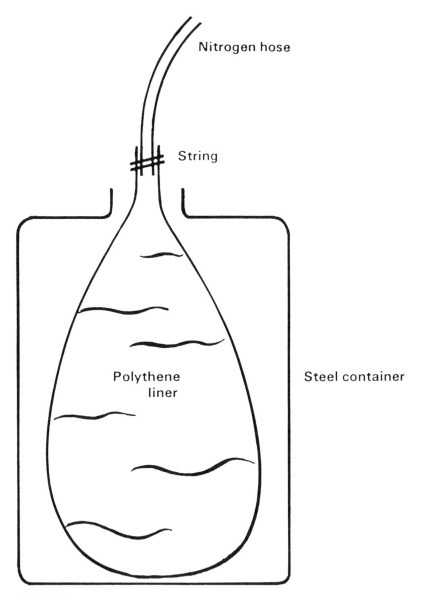

FIGURE 25.3 When the polythene linear was inflated, the steel container was bowed out.

One of the reactors developed a crack and had to be removed (for the reason see 5 below). It was replaced by a temporary 20-inch pipe, which had two bends in it to allow for the difference in height. The existing bellows were left in position at each end of the temporary pipe [Figure 25.4(b)].

The design of the pipe and support left much to be desired. The pipe merely rested on scaffolding. Because there was a bellows at each end it was free to rotate or squirm, and did so when the pressure rose a little above the normal level. This caused the bellows to fail [Figure 25.4(b)].

There was no professionally qualified engineer on the plant at the time the temporary pipe was built. The men who designed and built it—design is hardly the word because the only drawing was a full-scale sketch in chalk on the workshop floor—did not know how to design large pipes that are required to operate at high temperatures (150°C) and gauge pressures (150 psi or 10 bar), and made no attempt to think through the results of the modification. Very few engineers have the specialized knowledge to design highly stressed piping, but in addition the engineers at Flixborough did not know that design by experts was necessary or that modifications should be probed systematically. They did not know what they did not know (Fig. 25.5).

5. The crack in the Flixborough reactor was itself the result of a modification. There was a leak from the stirrer gland on the top of the reactor. To condense the leaking vapor, water was poured over the top of the reactor. Plant cooling water was used because it was conveniently available.

The water contained nitrates that caused stress corrosion cracking of the mild steel reactor. Nitrate-induced cracking was well known to metallurgists but was not well known to other engineers at the time.[5]

Water has often been poured over equipment to

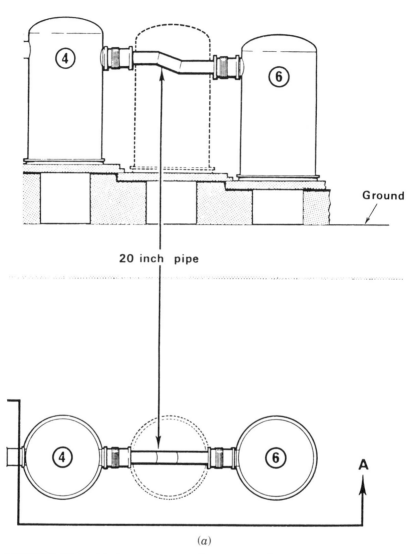

(a)

FIGURE 25.4 (a) Arrangement of reactors and temporary pipe at Flixborough. (b) Sketch of pipe and bellows assembly at Flixborough, showing shear forces on bellows and bending moments in pipe (due to internal pressure only). (Reproduced with the permission of the Controller of Her Majesty's Stationery Office).

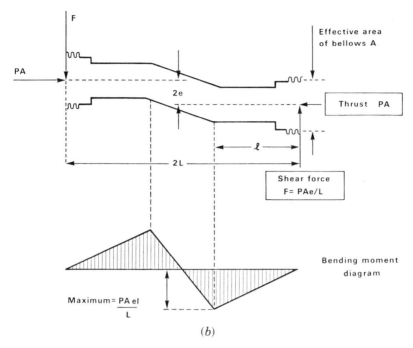

(b)

FIGURE 25.4 (continued)

cool it. Before doing so we should ask what is in the water and what effect it will have on the equipment.
6. In a wider field, Lawless[6] examined 46 technological changes, from thalidomide to plastic turf, that had unforeseen and undesirable side effects. He concluded that in 40% of the cases the side-effects could reasonably have been foreseen and that in 25% more notice might have been taken of early warning signs.

For other examples of modifications that went wrong and for an account of ways of controlling modifications see references 4 and 7–10.

Any modifications should be designed, constructed, tested and maintained to the same standards as the original plant.

FROM THE OFFICIAL REPORT ON THE EXPLOSION AT FLIXBOROUGH ON JUNE 1st 1974

FIGURE 25.5

References

1. Kletz, T.A. 1980. *Loss Prevention.* 13:1.
2. *The Flixborough Cyclohexane Disaster.* 1975. London: Her Majesty's Stationery Office.
3. Lees, F.P. 1980. *Loss Prevention in the Process Industries.* 1980. Appendix A1. Kent, England: Butterworths.
4. Kletz, T.A. 1988. *Learning from Accidents in Industry:* Chap. 8. Kent, England: Butterworths.
5. *Guide Notes on the Use of Stainless Steel in Chemical Process Plant.* 1978. p. 22. Rugby, England: Institution of Chemical Engineers.
6. Lawless, E.W. 1977. *Technology and Social Shock.* New Brunswick, NJ: Rutgers University Press.
7. Kletz, T.A. 1976. *Chem Engineering Progr.* 72(11):48.
8. Lees, F.P. 1980. *Loss Prevention in the Process Industries.* Chap. 21. Kent, England: Butterworths.
9. *Hazards of Plant Modifications.* Hazard Workshop Module 2. Rugby, England: Institution of Chemical Engineers.
10. Kletz, T.A. 1988. *What Went Wrong?—Case Histories of Process Plant Disasters,* 2nd. ed., Chap. 2. Houston: Gulf Publishing Co.

MYTH 26

Plants are made safer by adding on protective equipment.

The usual procedure in plant design is:

Design the plant.
Identify the hazards.
Add on protective equipment to control the hazards.

Typical protective equipment is:

Gas detectors
Emergency isolation valves
Trips and alarms
Relief valves and flarestacks
Steam and water curtains
Flame arrestors

Nitrogen blanketing
Fire protection equipment such as insulation and water
  sprays
Fire-fighting equipment

Slowly, however, the industry is coming to realize that an
alternative approach is possible: The best way of preventing a
large leak of hazardous material is to use a safer material in
its place, or to use less of the hazardous material, or to use it
at lower temperatures and pressures. "What You Don't Have
Can't Leak." Such plants are inherently safer and do not have
to be made safe by adding on protective equipment.

To use an analogy, if the meat of lions was good to eat,
farmers would find ways of farming lions. Cages and other
protective equipment would be required to keep them under
control and only occasionally, as at Flixborough or Bhopal,
would the lions break loose. But why keep lions when lambs
will do instead?

Reference 1 describes some of the changes that have been
or might be made, including the following:

1. Nitroglycerine used to be manufactured in a batch
   reactor containing 1 ton. Now it is made in a small
   continuous reactor and the residence time has been
   reduced from 120 minutes to 2 minutes.
2. Adipic acid used to be made in a reactor fitted with
   external coolers. Now it is made in an internally
   cooled reactor, thus eliminating pump, cooler, and
   pipelines. Mixing is achieved by the gas produced as a
   by-product.
3. The new Higee distillation process reduces the quantity of material in the distillation equipment by a factor of up to 1000. Even in conventional distillation
   equipment the inventory can be reduced by a factor of
   two or three by attention to detail.
4. By choice of suitable heat exchangers the inventory
   can be reduced by a factor of 20 or more.
5. Water can be used as a heat-transfer medium instead
   of flammable oils. If flammable oils must be used, high
   boiling point oils are available.

6. Fluorinated hydrocarbons can be used as refrigerants instead of hydrocarbons such as propylene.

Related to the concept of inherent safety is that of simpler plants. If we can spot hazards *early* in design we may be able to *remove* them by a change in design instead of *controlling* them by adding on protective equipment.

An example is provided by Section 1.3 and Figure 1.1 of Myth 1. We may be able to avoid the need for relief valves and the associated flare system by using stronger vessels. If we are to do so, the decision must be made early in design because vessels are normally ordered before detailed design is complete. It is no use waiting until the hazard and operability study or relief and blowdown review is carried out late in design. By this stage all we can do is to add on protective equipment, that is, relief valves, flare systems, etc.

Reference

1. Kletz, T.A. 1985. *Cheaper, Safer Plants or Wealth and Safety at Work—Notes on Inherently Safer and Simpler Plants.* 2nd ed. Rugby, England: Institution of Chemical Engineers.

MYTH 27

The technical people know their job. The safety adviser can leave technical problems to them and concern himself with human failings.

This myth was believed by many old-time safety advisers, partly because they were incapable of dealing with technical problems, but also because they genuinely believed that the technical staff had the technical problems under control. While this may have been true at one time, unfortunately it is not true today, as is shown by some of the incidents described earlier, for example:

The fire at Feyzin (Myth 2, Section 2.4).
Fires and explosions involving high boiling point materials (Myth 9).

The vessel burst by compressed air because the vent was choked (Myth 16).
The modifications described under Myth 25.

One of the jobs of the safety adviser in the process industries is to point out to his colleagues in design and production facts they ought to have learned early in their careers or, rather, to teach them to apply to practical situations the knowledge they have.

The safety adviser in the high-technology industries needs to be primarily[1]:

A technical man, with knowledge and experience of the technology of his industry, able to hold his own with managers and designers.
A numerate man, able to apply systematic and numerical techniques.
A communicator by speech, writing and discussion leading.
An auditor, spotting incipient trouble by sharp eyes and regular surveys.

In addition he should understand the limitations of men and know when it is reasonable to rely on men and when we should try to remove opportunities for error (see Myths 6–8).

Reference

1.   Kletz, T.A. 1982. *Safety and Loss Prevention.* Symposium Series No. 73, p. B1. Rugby, England: Institution of Chemical Engineers.

MYTH 28

The public believes we are making the world a better place.

If we believed this at one time, very few of us believe it today. Scientists and technologists are attacked by the media for producing hazards and pollution and, on the whole, we are not hitting back but waiting for the storm to blow over and wondering if the attacks are perhaps not partly justified.

Many have lost the self-confidence that scientists had before World War II, when it was taken for granted that science was the road to a better future.

This criticism of science and technology is not as new as is often thought. In "Victorian Engineering,"[1] L.T.C. Rolt wrote of the years following the Great Exhibition of 1851:

> The public began to lose confidence in the engineer so that he began to lose confidence in himself. The mood of uncritical reverence for material progress which had made the engineer the hero of the hour did not long outlive the Great Exhibition. The golden calf of the new Utopia, of man's emancipation by machine, was worshipped at the Crystal Place for the last time. A mood of doubt and disillusionment began, slowly at first, to undermine the old, easy confidence.

As an example of the new mood Rolt said that the Forth Bridge was attacked as "the supremest specimen of all ugliness."

The scientist is now suffering the loss of popular esteem that the engineer experienced 100 years earlier.

How Should We React?

Most of the opponents of science and technology are well meaning, but often muddled or ill informed. It is sometimes difficult to communicate with them because of a lack of common ground and because, when illogicalities and inconsistencies are pointed out, we may be told that "A wise judgement contains no logic."[2]

Nevertheless, we should not ignore the comments of the opponents of technology, but point out the facts and hope, like Lord Rothschild, "that if some course of action is made self-evident by hard information, and if that course of action is not followed, you and I will say to the politician who denies it what Queen Elizabeth the First said to her man Cecil: 'Get out!.' "[3]

Some of the facts that need to be pointed out are:

1. *We cannot have the benefits of modern technology without some disadvantages in terms of pollution and safety.*

   It is true that some of the benefits are ones of doubtful

value (striped toothpaste, for example), and the opponents of technology are quick to point this out. They are less ready to dispense with antiseptics and anesthetics.

2. *New technologies are usually less hazardous than old ones.*

   Flixborough has been described as "the prices of nylon," and many must have wondered if it is worth the risk, but we have to wear clothing of some sort, and the 'accident content' of natural fibers such as wool and cotton is higher than that of man-made fibers. (The cost of an article is the cost of its labor content, capital costs being other people's labor. Natural fibers 'contain' more labor because agriculture is a low-wage industry and they 'contain' more accidents because agriculture is a high-accident industry). Similarly, nuclear energy, despite Chernobyl, is safer than energy derived from oil or coal[4-6], plastics are safer to produce than metal or wood.

   One might have expected the environmental lobby to have favored nuclear energy because it is so much safer and cleaner than other types. Perhaps they associate nuclear energy with atomic bombs.

3. *The cost of reducing pollution and increasing safety has to be paid for in the end by the public.*

   There comes a time when even the most safety-conscious person objects to paying a lot of money for a further reduction in an already very small risk and prefers to use the money in other ways. The more we spend on safety the less there is left to spend on reducing poverty and disease or on those goods and services that make life worth living, for ourselves and other people.

4. *Men, not technology, create hazards and pollution.*

   T.C. Young wrote:

   A human society has a technology, and technology does not exist outside of the context of a human society. To see pollution as the result of technology and nature out of balance is to shrink from reducing the equation to its lowest common denominator. To blame pollution on technology is the ultimate dodge of a society unwilling to take the blame for its own errors and stupidity. We are so accustomed to excusing everything from misbehaviour to crime, from social dislikes to murderous hates, on man's social environment rather than on man, that when we come to deal with the natural environment we refuse from habit to take the blame on ourselves. Instead we pretend that technology, our technology, is something of a life force, a will, and a thrust of its own, on which we can blame all, with which we can explain all, and in the end by means of which we can excuse ourselves.

Not so. It is people who make pollution, not technology.[7]

They have always done so, and the defenders of the environment are often defending the results of man's past activities in earning a living. The English countryside, for example, bears no resemblance to its appearance before the coming of man and is as much a man-made feature as a factory.

Similarly, it is not computers and automation that cause unemployment but the way we use them.

References

1. Rolt, L.T.C. 1974. *Victorian Engineering.* pp. 162, 195. London: Penguin Books.
2. van Dieren, W. 1978. *Chemistry and Industry, 2,* December, p. 899.
3. Lord Rothschild. *Risk.* p. 13. London: BBC Publications.
4. Health and Safety Commission. 1978. *The Hazards of Conventional Sources of Energy.* London: Her Majesty's Stationery Office.
5. Inhaber, H. 1978. *Risk of Energy Production.* Report No AECB-1119/REV-1. Ottawa, Canada: Atomic Energy Control Board.
6. Cohen, A.V. and Pritchard, D.K. 1980. *Comparative Risks of Electricity Production. A Critical Survey of the Literature.* Research Paper 11, Health and Safety Executive. London: Her Majesty's Stationery Office.
7. Young, T.C. 1975 In *Man in Nature,* ed. L.D. Levine, Toronto, Canada: Royal Ontario Museum.

MYTH 29

Major failures of plant and equipment are now so infrequent that it is rarely possible to reduce them further.

No one ever actually says this, but people imply that they believe it by spending a great deal of money and effort estimating the *probability* of major failures and the *consequences* of these failures, but very little effort considering their *causes* and methods of prevention.

For example, in 1982 the Institution of Chemical Engineers organized a symposium on "The Assessment of Major Hazards."[1] The 24 papers presented divided about equally into those that discussed the *probability* of a major incident and those that discussed the *consequences*. Not one paper discussed the causes. It was as if the occasional major leak was considered inevitable.

At the Fourth International Symposium on Loss Prevention[2] held in England in 1983 over 100 papers were presented. Over half considered methods of estimating the probabilities and consequences of major leaks. About a dozen papers described incidents that have occurred and made specific recommendations to prevent them happening again, but there were no general papers on the causes of leaks.

Expensive experiments have been carried out to determine the behavior of major leaks—how they disperse in the atmosphere and how they behave when ignited.[3] Millions of pounds have been spent. In contrast, little effort has been devoted to studying the causes of leaks, though doing so would cost a minute fraction of the money spent on studying the behavior of leaks. Yet if we could prevent leaks we would not have to worry so much about the behavior of the leaking material.

Why has there been this imbalance in the allocation of resources? Do we prefer to tackle the more interesting problems—or the more fasionable ones—instead of those that most need tackling? This is often looked upon as an academic failing, but most of the experiments I have described have been carried out by industry—or paid for by them.

My own examination of reports on major leaks[4-6] suggests that the biggest single cause is pipe failure and that many of these failures arise at the design-construction interface—the construction team do not follow the design, or details are left to them and are not carried out in accordance with good engineering practice. To reduce the number of major leaks we should specify designs in detail and then check after construction to make sure that the design has been followed and that details not specified have been carried out in accordance with good engineering practice.

Here are some examples of accidents caused by a failure to follow designs in detail or to do well what has been left to the discretion of the construction team:

1. Small-bore branches have been inadequately supported, and have vibrated and failed by fatigue.
2. Old pipe has been re-used. It was already corroded or had used up some of its creep life (see MYTH 47).
3. Temporary supports have been left in position, or temporary branches, installed for pressure testing, not welded up.
4. The end of a relief valve tailpipe has been placed so close to the ground that it was sealed by a frozen puddle.
5. Dead-end branches have been left in pipework, and have filled with water (which froze, breaking the line) or with corrosive by-products.
6. Bellows have been installed between pipes that were not in line, so that the bellows were distorted.

References

1. *The Assessment of Major Hazards.* 1982. Symposium Series No. 71. Rugby, England: Institution of Chemical Engineers.
2. *Loss Prevention and Safety Promotion in the Process Industries.* 1983. Symposium Series Nos. 80–82. Rugby, England. Institution of Chemical Engineers.
3. Pikaar, M.J. 1983. In *Loss Prevention and Safety Promotion in*

the Process Industries. Symposium Series Nos. 80–82. Vol. 1, p. C20. Rugby, England: Institution of Chemical Engineers.

4. Kletz, T.A. 1980. *Safety aspects of pressurized systems.* In *Proceedings of the Fourth International Conference on Pressure Vessel Technology.* London: Institution of Mechanical Engineers, May. p. 25.
5. Kletz, T.A. 1983. *Plant/Operations Progr* 3(1):19.
6. Kletz, T.A. 1988. *Learning from Accidents in Industry.* Chap. 16. London: Butterworths.

MYTH 30

If equipment has to be opened up frequently then quick-release couplings should be fitted to save time and effort.

Every day, in every process factory, equipment that has been under pressure is opened up for cleaning or repair. Normally a process operator prepares the equipment, by blowing off the pressure and, if necessary, removing any traces of the contents that remain, and issues a permit-to-work to a maintenance worker who opens up the equipment, usually by unbolting the bolts that hold the cover or flanges in place. If there is any pressure left in the equipment, and the bolts are slackened correctly, this is immediately apparent and the bolts can be tightened or the pressure allowed to blow off.

Safety is achieved by (1) a procedure that makes one person responsible for preparation of the job and another responsible for actually carrying it out, and (2) an opening technique that can cope with a failure to carry out the preparation correctly.

When equipment has to be opened up frequently it is sometimes designed so that the entire operation can be carried out by the operator, using quick-release couplings. Sooner or later, by oversight or neglect, an attempt is made to open up the equipment before the pressure has been blown off. Several accidents that have occurred in this way are described below.

If quick-release couplings are used, therefore:

1. Interlocks should be provided so that the equipment cannot be opened up until the source of pressure is isolated and the vent valve opened, and/or

2. The design of the door or cover should allow it to be opened about ¼ inch (6 mm) while still capable of carrying the full pressure, and a separate operation should be required to release the cover fully. If the cover is released while the vessel is under pressure, then this is immediately apparent and the cover can be resealed and the pressure allowed to blow off through the gap.

Precaution 1 alone is not acceptable if the vent is liable to choke.

Accidents Resulting from the Opening of Pressure Vessels Using Quick-Release Couplings

1. During the 1960s, a suspended catalyst was removed from a process stream in a pressure filter. After filtration was complete, steam was used to blow the remaining liquid out of the filter. The pressure in the filter was blown off through a vent valve and the fall in pressure was observed on a pressure gauge. The operator then opened the filter for cleaning. The filter door was held closed by eight radial bars that fitted into U-bolts on the filter body. The bars were withdrawn from the U-bolts by turning a large wheel, fixed to the door. The door could then be withdrawn.

    One day an operator started to open the door before blowing off the pressure. He was standing in front of it and was crushed between the door and part of the structure, and was killed instantly. In this sort of situation, sooner or later, through oversight or neglect, an attempt will be made to open the equipment while it is under pressure; on this occasion the operator was at the end of his last shift before going on holiday. It is too simple to say that the accident was due to the operator's mistake. The accident was the result of a work situation that made an accident almost inevitable. However, this was not fully recognized at the time, and less change was made to the plant than we would make today[1] (see also Myth 16).

2. Plastic pallets were being blown out of a tank truck by compressed air. When the tank seemed empty, the driver opened a manhole cover on the top to make sure. One day a driver, not a regular, started to open the manhole before releasing the pressure. When he had opened two of the quick-release couplings, the cover was blown open. The driver was blown off the tank and killed by the fall.

   Either the driver forgot to vent the tanker or he thought that it would be safe to let the pressure (10 psig) blow off through the manhole. After the accident the manhole covers were replaced by the type described above, and the vent valve was moved from the side of the tanker to the foot of the ladder.[1]

   Many of those concerned were surprised that 10 psig could cause so much injury (see Myth 16).

3. An operator was using a portable sprayer that was pressurized by a plunger-type pump. The spray reduced to a dribble and the sprayer seemed light in weight, so he opened it to refill it. To do so he had to tap the screwed top loose with a hammer. The top came away with sufficient force to knock him off his feet. He had not used this type of sprayer before.

   Investigation showed that the reduced spray was due to a plugged nozzle. The sprayer was scrapped and replaced by one fitted with a two-stage opening device, as described above, and a pressure gauge.

4. On a new plant a large filter on a 14-inch liquid propane line was fitted with a process-operated filter. The filter was opened while still under pressure and two men were hospitalized with cold burns.

   The company concerned had agreed some years before that all pressure vessels that can be opened by operators should be fitted with two-stage opening devices. However, filters were classified as 'pipe fittings' rather than vessels, and were designed by the piping

FIGURE 30.1 Vacuum flask stopper. Any trapped pressure will be released before the stopper is unscrewed.

section, who were not familiar with the vessel specifications.

With small filters (containing, say, only a few kilograms of liquid) the precautions necessary with vessels may not be required, but they are necessary with large filters.

An Illustration from Another Branch of Technology

Figure 30.1 shows the stopper from a vacuum flask. It is hollow and there is an opening in the wall of the threaded section. The stopper has been made this way to reduce heat transfer, but it has another advantage. When the stopper is unscrewed a few turns, any trapped pressure is immediately apparent. It can be allowed to blow off or the stopper screwed back. The contents of the flask cannot be ejected.

Reference

1. Kletz, T.A. 1985. *An Engineer's View of Human Error.* Chap. 2 Rugby, England: Institution of Chemical Engineers.

MYTH 31

The transport of chemicals is dangerous.

At first sight, there seems to be a good deal of evidence for his statement. At San Carlos de la Rapita in Spain in 1978 over 200 people were killed when a tank truck of propane disintegrated and the contents ignited.[1,2] In Germany, in 1943, 209 people were killed when a rail tanker of ether burst and the contents exploded.[3] In New York state in 1958 200 people were injured when a rail tanker of nitromethane exploded.[4] More recently, 46 people were killed in the United States by accidents involving rail tankers during the period 1969–78[5] (see also Myth 32).

In contrast, in the United Kingdom the road transport of chemicals and gasoline kills an average of 1.2 people per year.[6,7] (Sixteen were killed in the period 1970–82, about half drivers, about half in accidents involving gasoline). This excludes accidents in which people were killed by impact and the contents of the tanker were not involved. So far as rail transport is concerned, I do not know of any accident in which people have been killed.

Have we just been lucky in the United Kingdom, or is our freedom from serious accidents due to higher standards? I think the latter is probably the case, particularly in the following respects:

1. In Europe types of steel are used for tank trucks that would not be used (for certain loads) in the United Kingdom. The tank involved in the Spanish disaster was made from T1 steel and had been used for carrying ammonia, and as a result stress corrosion cracking had occurred. A similar tank failed in France in 1968 while being filled with ammonia, and five people were killed.[8,9]

2. In the United Kingdom tank trucks carrying liquefied flammable gases under pressure are fitted with relief valves, but in Europe this is not the case. Relief valves would probably have prevented the German incident and many lesser incidents.

3. In the United States the railway track is in much poorer condition than in the United Kingdom, and derailments are more frequent. These have caused many tank accidents. Others have been due to the free shunting of tank cars, a practice not permitted in the United Kingdom. United States tank cars have now been fitted with improved couplers, with head shields to prevent couplers piercing the end of the tank cars, and with insulation to protect the tank cars from fire, and these measures have reduced the numbers of people killed.[5] No one seems to have asked if it would have been cheaper to improve the track.

References

1. Stinton, H.G. 1983. *J Hazardous Materials* 7:373. (There are a number of inaccuracies in this account.)
2. Hymes, I. 1983. *The Physiological and Pathological Effects of*

    Thermal Radiation. Report No SRD R 275. App. 5. Warrington, England: U.K. Atomic Energy Authority.

3. Lewis, D.J. 1983. *Hazardous Cargo Bull* 4(8):12.
4. Lewis, D.J. 1983. *Hazardous Cargo Bull* 4(9):36.
5. U.S. National Transportation Safety Board. 1979. *Safety Report on the Progress of Safety Modifications of Railroad Tank Cars Carrying Hazardous Materials*. Washington, D.C.: U.S. Government Printing Office.
6. Kletz, T.A. 1984. *Hazardous Cargo Bull* 5(11):10.
7. Kletz, T.A. 1986. *Plant/Operations Progr* 5(3):160.
8. *Annales des Mines*, January 1969, p. 25.
9. *Ammonia Plant Safety*, 1970, Vol. 12, p 12.

MYTH 32

Major disasters in the chemical industry are becoming more frequent.

This is the impression we get from the media, and it is supported by statements such as the following:

> The trends in frequency and proportion of incidents producing blast and fatalities in UVCE's (unconfined vapour cloud explosions) . . . are all upwards.[1]
> . . . the number of incidents is on the increase despite the activities of the past few years.[2]
> . . . there is little evidence that the oil crisis has checked the increase in rate of occurrence of serious incidents.[3]
> The impression was that the frequency of disasters on an international scale was increasing.[4]

On the other hand, another author wrote:

> By almost every type of societal indicator, except one, hazardous events have been increasing. . . . The one exception is the statistical record of hazard consequences . . . there frequently appears an enormous divergence between this record and the perception of hazards by scientists, the public and officialdom.[5]

During the 1970s I kept records of serious incidents in the oil and chemical industries that were reported in the press and in sources such as references 1 and 6. From these I extracted details of all incidents in which five or more people were killed. They are summarized in Table 32.1, which shows that the number of incidents occurring on fixed installations (plants, storage areas, pipelines) was roughly constant during the period covered (1970–80) despite the growth in the size of the industry, which more than doubled during the decade. For fixed installations, the average number of incidents was 7 per year with a range of 4–13. The average number of people killed was 147 per year with a range of 55–505.

Transport accidents are a different matter. For the period 1970–77, the average number of incidents was 3 per year with a range of 2–6. The average number of peopled killed was 57 per year with a range of 12–228. However, in 1978 there were 9 incidents killing 384 people and in 1979 there were 9 incidents killing 177 people. The year 1978 was dominated, of course, by the Spanish tank truck disaster, but even without this incident it would have been a black year for the transport of oils and chemicals, exceeded only by 1972 and

TABLE 32.1   Annual numbers of serious incidents and deaths caused by them in the oil and chemical industries

Year	Total		Fixed installations		Transport	
	No of incidents	No killed	No of incidents	No killed	No of incidents	No killed
1970	7	127	5	115	2	12
1971	6	545	4	505	2	40
1972	12	364	6	136	6	228
1973	10	140	8	114	2	26
1974	7	189	5	166	2	23
1975	10	114	8	68	2	46
1976	15	156	13	141	2	15
1977	11	123	6	55	5	68
1978	17	483	8	99	9	384
1979	17	324	8	147	9	177
1980	9	137	6	71	3	66
TOTALS (11 yrs)	121	2,702	77	1,617	44	1,085
Mean (per year)	11	246	7	147	4	97
Fatalities/ Incident	—	22	—	21	—	24

1979. It should be noted, however, that only one of the transport incidents occurred in the United Kingdom (in 1979).

My coverage of the press was not exhaustive but it got better in the later years and therefore would tend to exaggerate any growth in the number of incidents.

My definition of the oil and chemical industries was very broad. It included accidents caused by conventional explosives, people killed while collecting gasoline from a leaking tank truck, people poisoned by eating stolen seed corn, and explosions due to leaks from natural gas transmission lines. Details are in reference 7.

References

1. Gugan, K. 1978. *Unconfined Vapour Cloud Explosions.* p. 103. Rugby, England: Institution of Chemical Engineers.
2. *Processing,* April 1979, p. 25.
3. Marshall, V.C. Quoted in *Processing,* April 1979, p. 25.
4. Barrell, A.C. quoted in *The Guardian,* 24 April 1979.
5. Kates, R.W. 1978. *Managing Technological Hazard—Research Needs and Opportunities.* Boulder: University of Colorado.
6. Nash, J.R. 1977. *Darkest Hours.* New York: Pocket Books.
7. Turner, E., and Kletz, T.A. 1979. *Is the Number of Serious Accidents in the Oil and Chemical Industries Increasing?* London: Chemical Industries Association.

MYTH 33

It must be safe because we have done it this way for years without an accident.

'Practical people' often talk like this—experience, they say, shows that the operation has been carried out in a particular way for 20 years without an accident occurring. That proves it is safe.

What do we mean by safe? Is an accident in the 21st year acceptable? If it is not, then we have not proved that the operation is safe, because an accident might occur then.

Even if one accident in 20 years is acceptable, we have not even shown that on *average* the accident rate is less than one in 20 years. Suppose that *on average* one accident occurs in a period of time such as 20 years; then on average there is a 37% chance that no accidents will occur in this period of time. Suppose on average 2 accidents occur in a period of time, then on average there is a 14% chance that no accidents will occur in this period of time.

These figures are derived from the Poisson distribution, which shows that if the average number of random events that will occur in a given time is μ, then the probability P that n events will occur is given by:

No. of events = 0	1	2	n
Probability $= e^{-\mu}$	$e^{-\mu}\mu$	$e^{-\mu}\dfrac{\mu^2}{2!}$	$e^{-\mu}\dfrac{\mu^n}{n!}$

If $\mu = 1$ and $n = 0$, then $P = e^{-1} = 0.37$
If $\mu = 2$ and $n = 0$, then $P = e^{-2} = 0.14$

Putting it the other way round, if no failure has occurred in 20 years, what is the probability that the average failure rate is 1 in 10 years or less?

If the average failure rate is 1 in 10 years then the probability of no failure in 10 years is 0.37 and the probability of no failure in the following 10 year period is also 0.37. The probability of no failure in a 20-year period is $0.37 \times 0.37 = 0.14$. If there has been no failure in a 20-year period we can be 86% confident that the average failure rate is 1 in 10 years or less.

Let us look at some accidents that occurred because, on the basis of past experience, people assumed that it was safe to continue.

1. At the BASF plant at Oppau, Germany in 1921, explosives were used to break up storage piles of a 50:50 mixture of ammonium sulfate and ammonium nitrate. Two terrific explosions occurred, killing 430 people, including 50 members of the public, destroying the plant and 700 houses, and producing a crater 250 feet across and 50 feet deep.
   The operation had been carried out without mishap 16,000 times before the explosion occurred.[1]

2. While starting up a coker—a plant for making coke—
    an operator forgot to open a valve. As a result an ex-
    plosion occurred and a man was killed. The plant and
    another similar one were each started up every few
    days because they had to be shut down for emptying
    between batches, and there had been 6000 successful
    start-ups before the explosion occurred.[2]
3. Nitromethane was considered nonexplosive and safe
    to transport in rail tanks from 1940 until 1958 when
    two tanks exploded in separate incidents in the
    United States. Both occurred as the result of shunting.
    Although only two people were killed, many were in-
    jured, damage was extensive, and the second incident
    produced a crater 30 m across and 10 m deep.[3]
4. H.A.L. Fisher, describing the fall of Constantinople in
    1453, wrote, ". . . since the city had never been taken,
    a belief prevailed that it could not be taken. Cities,
    however, are not defended by beliefs but by material
    power."[4]

References

1. Lees, F.P. 1980. *Loss Prevention in the Process Industries*. Vol. 2,
    App. 3. Kent, England: Butterworths.
2. Vervalin, C.H. 1985. *Fire Protection Manual for Hydrocarbon
    Processing Plants*. 3rd ed. p. 95. Houston: Gulf Publishing Com-
    pany.
3. Lewis, D.J. 1983. *Hazardous Cargo Bull* 4(9):36.
4. Fisher, H.A.L. 1936. *A History of Europe*. p. 408. London:
    Edward Arnold.

MYTH 34

**If there were five accidents last year and seven the
year before then the accident rate is getting better.**

In industry we often see reports in which a factory is
praised for a small decrease in the numbers of accidents or
blamed for a small increase. Is the change real, or is it just
due to chance? Figures 34.1 and 34.2 may help us decide.[1,2]
Two cases are considered.

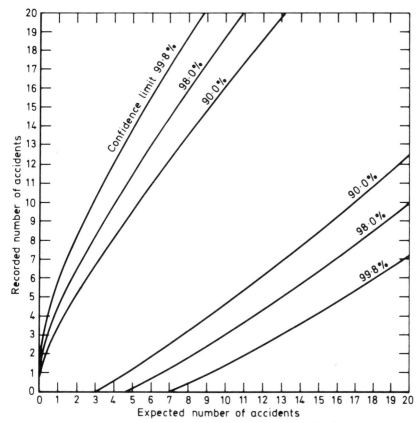

FIGURE 34.1 How to tell if the accident record is really better.

1. The Average (or Expected) Number of Incidents Is Known

Suppose that on average there are nine accidents (or nine leaks or nine fires) per month (or per week or per year). Suppose that in a particular month (or week or year) there are only four. Have things really gotten better?

Looking at 9 on the horizontal axis of Figure 34.1 and then following the vertical line upward, we see that it crosses the 90% confidence line at four. This means that we can be 90% certain that things have gotten better. If we say things have gotten better, nine times out of 10 we will be right.

If, however, there are five incidents one month, we ought to wait another month or two before congratulating anyone on an improvement.

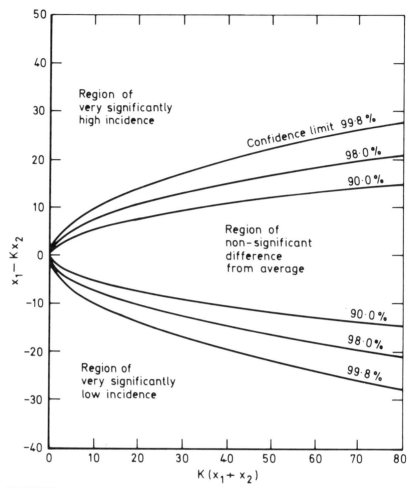

FIGURE 34.2   How to tell if the accident record is really better when the average record is not known.

If instead of four incidents there are only two, we can be 98% certain that things have gotten better, that is, 98 times out of 100 we will be right.

If there is only one incident, we can be 99.8% certain that things have gotten better; 998 times out of 1000 we will be right.

Similarly, if there are 15 incidents we can be 90% certain

that things are worse. If there are 17 incidents we can be 98% certain and if there are 20 incidents we can be 99.8% certain.

2. The Average Number of Incidents Is Not Known

So far we have assumed that we know the average (or expected) incident rate. Suppose there were 20 incidents in one period and 10 in the next and we do not know the average. Figure 34.2 can then be used. If:

$$x_1 = 20$$
$$x_2 = 10$$
$$x_1 + x_2 = 30$$
$$x_1 - x_2 = 10$$

find $x_1 + x_2$ on the bottom scale and move upward. We cross the 90% line at $x_1 - x_2 = 9$. $x_1 - x_2$ is actually equal to 10, so we are more than 90% certain (but less than 98% certain) that there has been an improvement.

In this example $K = 1$. K allows for the fact that in the two periods under comparison, the number of employees (or the number of hours worked or whatever we think is relevant) may not be the same. If there are N_1 employees (or hours worked) in the first period and N_2 in the second period, then $K = N_1/N_2$.

Let us now apply Figure 34.2 to two real examples:

1. A unit reported that there were seven lost time accidents in one year but only five in the next year. The staff were congratulated on a "commendable" improvement. However, if:

$$x_1 = 7$$
$$x_2 = 5$$
$$K = 1$$
$$x_1 + x_2 = 12$$
$$x_1 - x_2 = 2$$

from Figure 34.2 we see that if $x_1 + x_2 = 12$, $x_1 - x_2$ must be five before we can be 90% confident that an improvement has occurred. There is thus no evidence for a real improvement; a fall from seven to five could be due to chance.

2. A factory reported that there were 93 road accidents in one year but only 86 in the following year, though traffic had increased by 5%. Road users were congratulated on an "encouraging" improvement. However, if:

$$x_1 = 93$$
$$x_2 = 86$$
$$K = 0.95$$
$$K(x_1 - x_2) = 0.95 \times 179 = 170$$
$$x_1 - Kx_2 = 93 - 82 = 11$$

On Figure 34.2, 170 is off the scale, but using a similar graph we find that $(x_1 - Kx_2)$ must be 20 or less if we are to be 90% confident there has been an improvement. There is thus no evidence that the road accident record has really improved—the change may be due to chance.

Notes

1. The graphs cannot be used for comparing accident *rates;* they can only be used for comparing numbers of accidents (or other events).
2. Any conclusions drawn from Figure 34.2, will be wrong if the wrong value of K is used. For example, we might assume that the number of accidents depends on the number of hours worked and take K as the ratio between the number of hours in the first period and the number in the second. If the number of accidents is actually proportional to output, our conclusions may be wrong.

References

1. Wynn, A.H.A. 1950. *Applications of Probability Theory to the Study of Mining Accidents.* Report No. 7. Sheffield, England: Safety in Mines Research Establishment.
2. Lees, F.P. 1980. *Loss Prevention in the Process Industries.* Vol. 2. Sect. 27.5. Kent, England: Butterworths.

Appendix to Myth 34: The Manager's Dream

The following story shows how easily we can be misled by accident statistics—when the numbers involved are small.

It was an evening at the end of January 1984 and the Works Manager could put it off no longer. He had to get out his safety report for the previous year. This would show that his Works had had six lost-time accidents—the same as in the year before. Every other Works in the group had shown an improvement—between 10% and 40%—but his works had remained the same. It was bottom of the league and the Board had said that in 1983 the lost-time accident record was one of the things on which Works Managers would be assessed. It was rumored that the Works Manager showing the biggest improvement would be promoted, the one showing the worst . . . .

He gazed at the figures showing accidents month by month, hoping that somehow one of them would disappear.

	1982	1983	1984
Jan	0	2	0
Feb	0	0	
March	1	0	
April	0	1	
May	1	0	
June	0	1	
July	1	0	
Aug	0	1	
Sept	0	1	
Oct	2	0	
Nov	0	0	
Dec	1	0	
Total	6	6	

While the Works Manager was working his wife was watching television. He looked up to see the President announce that in future the year would start on February 1 instead of January 1.

Looking at his figures, the Works Manager saw that his

accident record showed a remarkable improvement—from 8 in 1982 to 4 in 1983—he had halved it.

	1982	1983	1984
Jan	0	2	0
Feb	0	0	
March	1	0	
April	0	1	
May	1	0	
June	0	1	
July	1	0	
Aug	0	1	
Sept	0	1	
Oct	2	0	
Nov	0	0	
Dec	1	0	
Total	6	6	
	8	4	

He always knew he could improve the safety record if he set his mind to it.

"Wake up dear," said his wife. "You've fallen asleep again."

All that the Works Manager had to do was find a convincing reason for starting the safety year in February rather than January. That should not be too difficult—the Safety Committee took office on February 1; the safety competitions started on February 1. They had not done so in the past, but that could soon be changed. As for next year—he looked forward to promotion and his successor could worry.

MYTH 35

The effectiveness of a relief valve is not affected by the position of the vessel.

Plant vessels do not usually move, but tankers may topple over as the result of an accident. The relief valve may then be below the liquid level so that it discharges liquid, not vapor. If the tanker contents are flammable the relief valve will have been sized to pass the volume of vapor produced when the tanker is surrounded by fire. To prevent the tanker being overpressured and possibly bursting, the relief valve would have to pass a volume of liquid equal to the volume of vapor produced and it is, of course, much too small to do this.

This cause of bursting is quite different from that which occurs in a BLEVE (see Myth 2) when the metal above the liquid level is softened by fire and loses its strength.

MYTH 36

An oversized relief valve will give increased safety.

Not always, because if the valve is far too big it may chatter and the vibration may cause a joint to leak. A serious fire that started in this way is described in reference 1. The vibration may be particularly severe if there are two relief valves and they compete with each other and cycle rapidly. If two relief valves have to be installed one should be set at a slightly lower pressure than the other.

Reference

1. Politz, F.C. 1985. *Poor Relief Valve Piping Design Results in Crude Unit Fire*. American Petroleum Institute 50th Mid-year Refining Meeting, Kansas City, Missouri.

MYTH 37

A few inches of water in the bottom of a heavy fuel oil tank will not cause any trouble.

It can cause trouble in two ways. If oil above 100°C is added to the tank the water will vaporize with explosive violence, the vent will probably not be big enough to pass the mixture of steam and oil, and the roof will be blown off the tank.

Alternatively, if the oil added to the tank is heated above 100°C, the heat will gradually travel through the oil to the water layer. When the water boils, the steam lifts up the oil, reducing the pressure and the water boils with increased vigor. Again the mixture of steam and oil will probably blow the roof off the tank. In one incident a structure 80 feet tall was covered with oil. Witnesses said that the tank had exploded and it was an explosion, but a physical explosion rather than a chemical one.

The phenomenon is known as foam-over, froth-over, boil-over, slop-over, or puking. The term "boil-over" is usually used if the tank is on fire and hot residues from the burning layer travel down to the water layer. The term "slop-over" is often used if water from fire hoses vaporizes as it enters a burning tank.[1]

To prevent foam-overs, which are very common in the oil industry, keep incoming oil *below* 100°C. It is good practice to fit a high-temperature alarm on the oil inlet line.

Alternatively, keep the tank contents well *above* 100°C so that any small quantities of water that enter are vaporized and a water layer never builds up. In addition the tank should be drained regularly and circulated before any fresh oil is added. Addition of fresh oil should start at a low rate and not more than a tenth of the tank capacity should be added at a time.

The temperature of a heavy oil tank should never be allowed to fluctuate above and below 100°C.

Reference

1. *Loss Prevention Bull,* No. 57, June 1984, p. 26; No. 59, October 1984, p. 35; and No. 61, February 1985, p. 36.

Further Reading

Kletz, T.A. 1988. *What Went Wrong?—Case Histories of Process Plant Disasters,* 2nd ed. Sect. 9.1.1, 12.2, & 12.4.5. Houston: Gulf Publishing Co.

MYTH 38

If you can seen the bottom of a hole in the ground it is safe to go into it.

Unfortunately not. Many men have been overcome and killed because they entered pits or depressions or other confined spaces without adequate testing. A tombstone in St. Cuthbert's Church, Marton, Middlesbrough, England reads:

Erected in memory of Robert Armstrong aged 28, James Ingledew aged 39 and Joseph Fenison aged 27 years who unfortunately lost their lives on Oct 11th 1812 by venturing into a well at Marton when it was filled with carbonic acid gas or fixed air. From this unhappy incident let others take warning not to venture into wells without first trying whether a candle will burn in them; if the candle burns to the bottom they may enter with safety; if it goes out human life cannot be supported.

According to the parish magazine the three men entered the well to recover some stolen beef they had hidden there.

A more recent incident[1] is typical of many. Some diesel fuel tanks were located in a pit. A man went down a ladder to drain water from the tanks. The job was done regularly but there were no instructions, no tests, and no breathing apparatus. The man collapsed. Another man entered to rescue him and also collapsed. The first man recovered but the second one died.

In the United Kingdom the Factories Act, Section 30, lays down the precautions necessary before anyone enters a confined space in which there may be dangerous vapor or fume. The confined space must be tested and isolated from all sources of danger. If necessary breathing apparatus must be worn. Entry must be authorized in writing by an authorized person.

In 1984 there was an explosion in a water pumping station at Abbeystead, England, in which 16 people were killed. No one knew that methane might be present and so no precautions were taken.[2] However, it is surprising that people were allowed to enter an underground pumphouse without the atmosphere being tested for carbon dioxide.

References

1. *Health and Safety at Work,* 1984 (April). p. 10.
2. Health and Safety Executive. 1985. *The Abbeystead Explosion.* London: Her Majesty's Stationery Office.

MYTH 39

Fire is worse than smoke.

Sometimes it is. Fire causes more damage than smoke and flash fires and BLEVEs (see Myth 2) kill people before the smoke has had time to affect them. However, when fires occur in buildings, aircraft, and trains more people are killed by carbon monoxide and other toxic chemicals in the smoke than by the fire itself. Several companies are now producing smoke hoods that are designed to let people leave smoke-filled buildings and aircraft in safety. However, some of those that are available do not provide protection against carbon monoxide. At the time of this writing Germany is the only country with a standard and approval system for the hoods.

MYTH 40

Endothermic reactions are safe because they cannot run away.

In endothermic reactions the absorption of energy tends to reduce the temperature, and heat usually has to be supplied to the reactor to keep the reaction going so, correct, endothermic reactions cannot run away. However, this does not mean that they are safe because the products of endothermic reactions have a high bond energy and tend to be unstable.

In addition, with endothermic reactions, as with exothermic reactions, a deliberate or unwitting change in reaction conditions can result in a different reaction taking place, and this reaction may be violent.[1]

Reference

1. Bretherick, L. 1987. In *International Symposium on Preventing Major Chemical Accidents, Washington D.C., February 3–5.* p. 4.1. New York: American Institute of Chemical Engineers.

MYTH 41

Covering a spillage with foam will reduce evaporation.

Not always, because water draining from the foam may increase the rate of evaporation from a cold liquid such as chlorine or LP gas. To estimate the evaporation produced in this way, Harris[1] assumed that a quarter of the water in the foam would drain in the first 15 minutes and quoted a specimen calculation.

The higher the expansion of the foam the less the drainage, but high-expansion foams are light and may be blown away by the wind.

If covering a spillage with foam will increase evaporation then Harris suggests that it should be covered with a plastic sheet. A chlorine spillage could be covered by men wearing breathing apparatus, but I would not like to ask anyone to spread a plastic sheet over an LP gas spillage.

Reference

1. Harris, N.C. 1987. In *International Symposium on Preventing Major Chemical Accidents, Washington, D.C., February 3–5.* p. 3.139. New York: American Institute of Chemical Engineers.

MYTH 42

Better late than never.

Not always. Suppose one liquid is added slowly to another while the mixture is stirred or circulated. If the stirrer or circulating pump stops the two liquids may form separate layers. If mixing is then started again the heat of reaction may be more than the cooling system can handle and a runaway may occur. This has happened a number of times.[1,2]

A pipeline was wrapped after corrosion started. This made the corrosion worse and the pipeline failed catastrophically. If it had been left unwrapped it would probably have leaked before failure.

References

1. Kletz, T.A. 1988. *What Went Wrong?—Case Histories of Process Plant Disasters.* 2nd ed., Sect. 3.2.8. Houston: Gulf Publishing Co.
2. *Loss Prevention Bull,* No. 29, October 1981, p. 124; and No. 78, December 1987, p. 26.

MYTH 43

Grounding clips should be fitted across flanged joints.

Visiting plants I often see these clips fitted across flanged joints. They are installed to make sure that the equipment is grounded, so that a charge of static electricity will not accumulate on it.

Certainly, when flammable liquids or gases are handled all conducting equipment should be grounded, but it is not necessary to install grounding clips across joints. The bolts used to fasten the flanges together make good enough contact to ensure a ground connection, as anyone can easily check for themselves with a resistance meter. To prevent a charge of static electricity accumulating the resistance to ground should be less than 1 MΩ, but to allow a lightening discharge to flow to ground the resistance should be less than 7 Ω.

MYTH 44

Because water is incompressible hydraulic pressure tests are safe. If the vessel fails the bits will not fly very far.

Hydraulic pressure testing is safer than pneumatic testing, because much less energy is released if the equipment fails. Nevertheless some spectacular failures have occurred during hydraulic tests. In 1965 a large pressure vessel (16 m long by 1.7 m diameter), designed for operation at a gauge pressure of 350 bar, failed during a pressure test at the manufacturers. The failure, which was of the brittle type, occurred at a gauge pressure of 345 bar and four large pieces were flung from the vessel. One piece weighing 2 tons went through the workshop wall and traveled nearly 50 m. Fortunately there was only one minor casualty. The failure occurred during the winter, and the report recommends that pressure tests should be carried out above the ductile-brittle transition temperature for the grade of steel used. It also states that the vessel was stress-relieved at too low a temperature.[1]

Another similar failure has been reported recently; in this case substandard repairs and modifications were partly responsible.[2]

When carrying out pressure tests, remember that the equipment may fail and take precautious accordingly. If we were sure that the equipment would not fail, we would not need to test it. Remember also that if the temperature is too low equipment may fail during pressurization for service.[2] I do not know of any vessels that have failed in this way but bursting discs have ruptured because they were too cold.

References

1. *Br Welding Res Assoc Bull,* Vol. 7, Part 6, June 1966, p. 149.
2. Snyder, P.G. 1988. *Loss Prevention* 7(3):148.

MYTH 45

Electric heaters have a high efficiency.

It depends what we mean by efficiency. Almost all the electric energy used in an electric heater is converted into heat, but if the same amount of electricity is used to drive a heat pump, we get more useful heat. The electric heater has a high First Law of Thermodynamics efficiency (almost 100%) but a low Second Law efficiency (about 25% for a domestic heating system; that is, we could get about four times more heat if we used the electricity to drive a heat pump).[1]

There is a good account of heat pumps in a book by Sumner.[2] Although everyone has one or two in their kitchen, many engineers are surprisingly ignorant of their capabilities. Sumner quotes a lecturer who said that a heat pump is impossible because it offends thermodynamic laws!

According to Sumner, the electricity industry in the United Kingdom has discouraged the development of heat pumps because they would reduce the demand for electricity.

References
1. Linnhoff, B. 1983. *Chem Engineering Res Design* 61(4):207.
2. Sumner, J.A. 1976. *Domestic Heat Pumps.* Dorchester, England: Prism Press.

MYTH 46

Accidents are due to a coincidence of unlikely events.

We are often told, after an accident, that it occurred because several safety devices failed simultaneously, an unlikely coincidence that could not reasonably have been foreseen, so no one is to blame. In fact, what usually happens is that all the safety devices are left for long periods of time in a failed state. When a triggering event occurs the accident is inevitable. For example, when an oil tanker was struck by lightning the vapor coming out of the tank vents was set alight and the flame traveled back along the vent line into the tanks, where an explosion occurred, killing two men. Three independent safety systems were each capable of preventing the explosion:

1. Nitrogen blanketing—it was not being used.
2. A flame trap—it was fitted incorrectly, with a gap round the edge.
3. A pressure/vacuum valve. This opens only momentarily to discharge vapor and the flow rate is too high for the flame to travel back, but a bypass round the valve had been left open.[1]

There was no coincidence of instantaneous events. There were three ongoing unrevealed faults, which stayed unrevealed because there were no regular tests or inspections. When lightning occurred an explosion was inevitable.

Reference

1. *Hazardous Cargo Bull,* Vol. 4, No. 7, July 1983, p. 8.

MYTH 47

We can save money by re-using old pipe, etc.

If we are altering or extending the plant it seems sensible to re-use old pipe that shows no signs of corrosion and looks fit for re-use. Unfortunately it is not, because the pipe may be affected in ways that are hard to detect even by expert examination. If it has been used hot, it may have used up some of its creep life (see Myth 21) and may fail in service under conditions in which new pipe would not fail. If the old pipe is made from stainless steel and has been in contact with chloride it may have been affected by stress corrosion cracking. The amount of chloride in town water may be sufficient. Old pipe should never be re-used unless we know its history and can take advice from a materials specialist.

There are many other examples of accidents caused by penny-pinching. The piston of a reciprocating engine was secured to the piston rod by a nut, which was locked in position by a tab washer. When the compressor was overhauled the tightness of this nut was checked. To do this the tab on the washer had to be knocked down and then knocked up again. This weakened the washer so that the tab snapped off in service, the nut worked loose, and the piston hit the end of the cylinder, fracturing the piston rod.

The load on a 30-ton hoist slipped, fortunately without injuring anyone. It was then found that a fulcrum pin in the break mechanism had worked loose because the split pin holding it in position had fractured and fallen. The bits of the pin were found on the floor.

Split pins and tab washers should not be re-used but replaced every time they are disturbed. Perhaps it is not penny-pinching but lack of spares that prevents us doing so. Perhaps we cannot be bothered to go to the store. Perhaps there are none in the store.

To quote a Chinese proverb, "If you go to bed early to save candles, the result may be twins."

2

Myths about Management

Engineers should respond to market needs.

So they should. If our marketing colleagues see a growing market for mousetraps, engineers are asked to design a plant to make mousetraps, or to increase the output from an existing plant. Also, the engineers are expected to produce the mousetraps as cheaply and efficiently as they can.

However, this is not the whole story. Very often our marketing colleagues and the public do not know what they want until engineers tell them what they can have.

We did not know that we wanted home computers and video recorders until we were offered them. Earlier generations did not know they wanted TV sets or radios or telephones or motor cars. The operating management of British Rail did not know they wanted a 125-mph train until it was offered to them.[1] Engineers should take the initiative and put forward ways of improving the business. As Boeing says, "Planes make markets."

Writing about the history of ICI, the United Kingdom's largest chemical company, Carol Kennedy says, "Winnington, where Ludwig Mond had set up his ammonia-soda works in partnership with John Tomlinson Brunner in 1873, was steeped in the Mond Philosophy that research paved the way for industrial progress rather than have it serve needs already perceived by the market." She quotes Ludwig Mond as saying that "the inventor can create new wants," and a present-day ICI researcher, Derek Birchall, who argues in favor of 'science push' rather than 'market pull,' as saying, "The huge market for polyethylene (a Winnington invention) came about because it was discovered, and was not created because it was needed. . . ." Kennedy adds that this view is not popular today and "strikes a distinctly iconclastic note."[2]

However, chemical engineers are probably less able to invent new products than other engineers because the new products of the process industries are usually invented by chemists. However, if chemical engineers cannot invent new products they can find better ways of making products: new designs of reactors, separation equipment, heat transfer equipment, etc. A discussion of examples is beyond the scope of this book, but there are some examples of what has been and might be done in reference 3.

Reactors, separation equipment, and the like are, of course, the products of the process equipment industry, and reference 4 discusses the reasons why it has been so poor at innovation. One reason is the low level of employment of first-rate engineers, but university syllabuses are also criticized.

If chemical engineers have not always been so good as they might be at innovation, perhaps one reason lies in the next Myth.

References

1. *Modern Railways,* Vol. 41, No. 427, April 1984, p. 169.
2. Kennedy, C. 1986. *ICI—The Company That Changed Our Lives.* pp. 60, 181. London: Hutchinson.
3. Kletz, T.A. 1985. *Cheaper, Safer Plants or Wealth and Safety at Work.* 2nd ed. Rugby, England: Institution of Chemical Engineers.

4. Solbett, J. 1983. *Process Plant R & D and Innovation*. London: Process Plant Economic Development Committee of NEDO. (For a summary see *Chem Engineer,* No. 394, July 1983, pp. 5, 7.)

MYTH 49

The job of the engineer is to answer questions he is asked, and solve problems he is asked to solve.

Engineers are good at answering questions. Tell them your problem and before long you will have many solutions to choose between. We are less good at asking questions. We often ask the wrong question. Before answering a question we should ask ourselves what the questioner really wants to know and if he is asking the right question.

Here are some examples:

1. *Question:* How can we reduce the number of fires and explosions an our plants and the damage they cause?
    Answer: By detecting leaks automatically, isolating them by means of remotely operated valves, dispersing the leaking material by the use of open construction and steam and water curtains, removing sources of ignition, and finally, if the leak should ignite, by installing fire protection and fire-fighting facilities.
    Comment: Much less attention has been given to the question "Why do leaks occur and how can we prevent them?" The usual answer is by good design, construction, operation, and maintenance. I suspect that the most effective way will be by providing better control of construction and better inspection after construction (see Myth 29). More fundamentally, can we use nonflammable or less flammable liquids or use less of the flammable ones or use the flammable ones at lower temperatures and pressures? (See *Cheaper, Safer Plants or Wealth and Safety at Work.*[1])

2. *Question:* How can I improve the control of this process?

Answer: By adding on lots of control instrumentation.
Comment: Should we consider redesigning the process
so that it does not need to be controlled so accurately,
so that variations in plant conditions produce smaller
variations in performance?

3. *Question:* Many accidents have occurred because operators opened the wrong valve or forgot to open a valve.
How can we prevent people from making such a mistake?
Answer: By telling operators to be more careful, pe-
nalizing those who make mistakes, etc.
Comment: Men carrying out a routine task will make
occasional mistakes even though they are well
trained, well motivated, and physically and mentally
capable. Once we realize this, and if the occasional
mistakes cannot be accepted, then the problem becomes one of removing or reducing the opportunities
for error (see Myth 6).

4. *Question:* What research should we do on safety?
Answer: Many suggestions have been made, some of
them quite expensive; for example, large-scale tests
on gas dispersion and unconfined vapor cloud explosions.
Comment: While there are a lot of things we would
like to know, on the whole accidents are not the result
of lack of knowledge. Accidents occur because the
knowledge of how to prevent them, though well
known, is not known to the people concerned or they
know what to do but lack the will to get on with it.

5. *Question:* How can fire engines get to a fire more
quickly?
Answer: By providing devices for starting engines
more quickly, opening fire station doors automatically, setting traffic lights in favour of the fire engine,
and so on.
Comment: The problem was seen as an engineering
one. But the time between someone dialing 911 and
the fire station getting the message can be as long as
the journey to the fire. Little or no thought has been
given to ways of speeding up this step.[2]

6. *Question:* The level of carbon monoxide in a covered car park was too high so the question was asked, "How can the carbon monoxide be removed?"

    Answer: A scheme was prepared for forced ventilation. It was expensive and there was no guarantee that it would be effective.

    Comment: When someone asked if the formation of carbon monoxide could be prevented, it was found that the high level was due to cars driving round and round looking for somewhere to park. Traffic lights controlled by traffic counters were installed at the entrance.[3]

7. *Question:* The Negev desert in Israel is irrigated with water transported great distances by pipeline. Water has been obtained from deep (over 500 m) wells but it contains over 500 ppm chloride. Can this be removed so that the water is suitable for irrigation?

    Answer: Techniques are available.

    Comment: Changes in methods of irrigation (to the drip method) and treatment of the soil have allowed the water to be used without purification.

8. About 700 BC King Hezikiah built a tunnel (now open to the public) to bring water from a spring outside the walls of Jerusalem to a point inside the walls. This enabled him to withstand a siege by the Assyrians.[4]

    Archaeologists have been puzzled by the fact that the roof of the tunnel varies in height (from 1.5 m to 5 m). Instead of asking why the roof is so high in places, someone asked why the floor was so low. A possible reason then became apparent: when the tunnel was constructed the floor was too high in places for the water to flow and had to be lowered.

Why are we so much better at answering questions than at asking the right questions? Is it because we are trained at school and university to answer questions that others have asked? If so, should we be trained to ask questions?

References

1. Kletz, T.A. 1985. *Cheaper, Safer Plants or Wealth and Safety at*

Work. 2nd ed. Rugby, England: Institution of Chemical Engineers.

2. *Fire Prevention,* No. 145, 1981, p. 31.
3. *Health and Safety at Work,* October 1981, p. 94.
4. II Chron. 32:2–4, 30.

MYTH 50

Many questions can be answered by substituting numbers in well-established equations.

Below are two cases showing how the wrong answer was obtained in this way. The people who used the equations did so mechanically without comprehending the physical realities underlying them.

Case 1: Fractional Dead Times

We start with some definitions:

Hazard rate is the rate at which hazards occur; for example, the rate at which the pressure in a vessel exceeds the design pressure or the rate at which the level in a vessel gets too high and the vessel overflows.

A *protective system* is a device installed to prevent the hazard from occurring; for example, a relief valve or a high-level trip.

Tests are (or should be) carried out at regular intervals to determine whether or not each protective system is inactive or 'dead.' It is assumed that if it is found to be dead it is promptly repaired. The time between tests is the *test interval (T)*.

Demand rate (D) is the rate at which a protective system is called on to act; for example, the rate at which the pressure rises to the relief valve set pressure or the rate at which a level rises to the set point of the high-level trip. 'Demand' is used in the French sense (demander = to ask).

Failure rate (f) is the rate at which a protective system develops faults that prevent it from operating.

Fractional dead time is the fraction of the time that a protective system is inactive; i.e., it is the probability that it will fail to operate when required.

If the protective system never failed to operate when required, then the hazard rate would be 0.

If there was no protective system, then the hazard rate would be equal to the demand rate.

Usually the protective system is inactive or dead for a (small) fraction of the time.

A hazard results when a demand occurs during a dead period. Hence:

Hazard rate = Demand rate × Fractional dead time

To derive the fractional dead time we assume that failure is random. It will therefore occur, on average, halfway between tests and the protective system will be dead for half the test interval ($\frac{1}{2} T$). Failure occurs on average after an interval of $1/f$, so the fractional dead time is $\frac{1}{2} fT$ and the hazard rate $\frac{1}{2} DfT$.

An example may make this clearer. Consider a high level trip for which:

The failure rate f = 0.5/year (once in 2 years)
The test interval T = 0.1 year (5 weeks)
The demand rate D = 1/year

Failure will occur on average halfway between tests, so the trip will be dead for 2.5 weeks ($\frac{1}{2} T$) every 2 years ($1/f$). The fractional dead time is therefore 2.5/(2 × 52) = 0.024 and the demand rate is 1 × 0.024 = 0.024/year, or once in 42 years.

So far, so good. Now consider another example. Let:

The failure rate f = 0.5/year (as above)
The test interval T = 0.1/year (as above)
The demand rate D = 100/year

Substituting in the formula hazard rate = $\frac{1}{2} DfT$, we get

$$\text{Hazard rate} = \frac{1}{2} \times 100 \times 0.5 \times 0.1$$
$$= 2.5/\text{year}$$

However, this answer is a nonsense. The trip will fail once every 2 years. This failure will probably be followed by a demand rather than a test (as there are 10 demands but only one test in each 5-week period), the vessel will overflow, and the fault will be discovered and repaired. The hazard rate will be almost the same as the failure rate, 0.5/year. Thus 2.5/year would be the correct answer, if when the vessel overflowed, we allowed it to overflow twice per week until the next test was due.

If you find this hard to follow, consider the brakes on a car. Let

Failure rate f = 0.1/year (once in 10 years)
Test interval T = 1 year (as required by law)
Demand rate D = 10 000/year (a guess)

Substituting in the formula hazard rate = $\frac{1}{2} DfT$ we get

Hazard rate = $\frac{1}{2}$ × 10 000 × 0.1 × 1 = 500/year

Not even the worst drivers have this many accidents. Clearly the simple formula hazard rate = $\frac{1}{2} DfT$, which we derived, does not always apply. By using it mechanically, without thinking of the reality behind it, we got wrong answers.

A more correct formula is:

$$\text{Hazard rate} = f(1 - e^{-DT/2})$$

When $DT/2$ is small, this becomes

$$\text{Hazard rate} = \frac{1}{2} fDT$$

When $DT/2$ is large, this becomes

$$\text{Hazard rate} = f$$

For a table comparing these equations see reference 1.

Case 2: Factorial Costing

It was decided during the design of a new plant (total cost about $10M) that the control building should be strengthened so that it could withstand the effects of an unconfined vapor cloud explosion. The project manager later reported that the increase in cost was $100,000, a sum so large that some people questioned the wisdom of the decision, and an investigation was requested.

The civil engineer who had designed the control building said that the increase in cost was only $30,000, about 14% of the original building cost. The project manager had multiplied this figure by a factor of 3.3 to arrive at an overall cost increase of $100,000.

Factorial costing is a well-recognized method of estimating plant costs. The cost of the 'main plant items' (reactors, distillation columns, heat exchangers, etc.) is multiplied by a factor, usually in the range of 3–6 depending on the type of plant, to allow for land, foundations, structures, piping, electrical equipment, instruments, ancillary buildings, and utilities as well as design and construction. The control building was considered a main plant item and a factor of 3.3 was the achieved figure for that particular plant.

However, increasing the strength of the control room may have added a little to design and construction costs but would not have increased the other items listed above. The figure of $100,000 was a nonsense. It had been obtained by using a formula mechanically without thinking of the reality behind it or asking if it applied in this case.

My examples have been relatively simple ones but they are supported by Linnhoff, who wrote, ". . . the basic understanding of important principles . . . can be a far more powerful asset in design than systematic methods. Systematic methods exclude the engineer from the design task. Above all, design is a creative process and to exclude the engineer from it must be a mistake."[2] Linnhoff goes on to say that he has shifted his emphasis from the "invention of methods" to the "demonstration of principles in use."

Just as systematic methods exclude the engineer from the design task, overformal management systems may exclude the manager from the management task. For example, if a company has a rigid system for investigating accidents and informing people of the results, managers may be lured into a belief that all they need do is follow the system. In fact, each accident is different and a flexible approach is needed.[3] The system inhibits flexibility and innovation. Similarly, a system of detailed safety laws and regulations encourages managers to think that all they need do is follow the rules (see Myth 51).

Perhaps we can go further. In life as a whole there are (according to Isaiah Berlin) "hedgehogs" who have an all-embracing system of belief and conduct and "foxes" who have no such unifying vision. When the hedgehog is challenged he rolls up into a ball and displays his spikes— the rules that tell him what to do—the same strategy for all occasions. The fox is more flexible; he will vary his strategy to suit the occasion and will accept and seek to justify some degree of contradiction in aims and methods.

In safety, hedgehogs like a set of detailed regulations covering all eventualities; foxes prefer to do what is reasonably practicable (see Myth 51).

The hedgehog and fox metaphors are taken from the Greek poet Archilochus, who wrote, "The fox knows many things but the hedgehog knows one big thing."

References

1. Kletz, T.A. 1986. *Hazop and Hazan—Notes on the Identification and Assessment of Hazards*. 2nd ed. p. 46. Rugby. England: Institution of Chemical Engineers.
2. Linnhoff, B. 1983. *Chem Engineering Res Design* 61(1):207. 1, July, p. 207.
3. Kletz, T.A. 1988. *Learning from Accidents in Industry*, Kent, England: Butterworths.

MYTH 51

Accidents can be prevented by detailed rules and regulations.

Is this the best way of preventing industrial accidents? The United States and United Kingdom take different approaches. In the United States, and many other countries, the government writes a book of regulations that looks like a telephone directory, but is less interesting to read. These regulations have to be followed to the letter, whether or not they will actually make the plant safer or are appropriate to our particular problems.

In contrast, in the United Kingdom, under the Health and Safety at Work Act (1974), there is a general obligation on employers to provide a safe plant or system of work and adequate instruction, training, and supervision. It is up to the employer to decide what he considers to be a safe plant, etc, but if the Factory Inspector does not agree he will say so, and, if necessary, issue an Improvement or Prohibition Notice.

If there is a generally accepted code of practice, failure to follow it is *prima facie* evidence of an unsafe plant or system of work, but the employer can argue that the code is inapplicable in his case or that he is doing something as safe or safer.

The points in favour of the U.K. system are:

1. Codes are more flexible than regulations. They can be changed when new problems arise or new solutions are found to old problems, and in any case, as already stated, they do not have to be obeyed to the letter. Changing regulations, however, takes a long time (see example below).

2. Employers cannot shelter behind the regulations. They cannot say (as I have heard people say in some countries) "My plant must be safe because I have fol-

lowed all the regulations." (There was something of this attitude at Three Mile Island.[1]) They must look out for and control any hazards missed by the regulations.

3. It is impracticable to write detailed regulations for a complex, rapidly changing technology.

4. The U.K. system is a more powerful weapon in the hands of a Factory Inspector than a volume of detailed regulations. Under a regulatory system the Inspector has to show that a particular regulation has been broken. If not, he is powerless. Under the U.K. system it is sufficient to show that the employer has not provided a safe plant or system of work, so far as is reasonably practicable.

Some people think that under the U.K. system the employer has a soft option. Not so; the pressures on him are greater than under the old (pre-1974) system.

Here are two examples of the nonsenses that can arise under a system of detailed regulations that have to be observed to the letter.

1. Under the U.K. Factories Act (1961) and earlier legislation steam boilers must be inspected at intervals not greater than 26 months. Those who drew up the regulations had fired boilers in mind but the regulations apply to all boilers, including waste heat boilers on oil and chemical plants, which may be subjected to much less severe conditions than fired boilers. Shutting down plants in order to carry out the statutory inspection of a waste heat boiler may be expensive and unnecessary.

The Factory Inspectorate have power to grant exemptions but the procedure is cumbersome.

Other plant vessels (except air receivers and steam receivers) do not have to be inspected under the Factories Act. Regular inspection is necessary under the

Health and Safety at Work Act, because otherwise there would not be a safe system at work, but an appropriate inspection frequency can be chosen.

2. Under the U.K. Petroleum (Consolidation) Act 1928 tank trucks carrying petroleum spirit had to be emptied by gravity. They could not be emptied by blowing the contents out with compressed nitrogen.

Expecting a change in the law, a company ordered some tanks that were to be emptied in this way, for carrying petroleum spirit between two factories. In order to withstand the nitrogen pressure they had to be stronger than ordinary tanks and were therefore less likely to be damaged in an accident. The change in the law was postponed for several years and the Company had to decide what to do with the tanks. The Health and Safety Executive publicly said that they would turn a blind eye to their use but that it was possible, though unlikely, that the police might prosecute for a technical breach of the law, if one of the vehicles was involved in an accident. The company lawyer advised that the vehicles should not be used.

Reference

1. Kletz, T.A. *Hydrocarbon Processing* 61(6):187.

Further Reading

Report of the Committee on Health and Safety at Work (Chairman, Lord Robens). 1972. London: Her Majesty's Stationery Office. (For the background to the Health and Safety at Work Act.)

Her Majesty's Inspector of Factories 1833–1983: Essays to Commemorate 150 Years of Health and Safety Inspection. 1983. London: Her Majesty's Stationery Office. (Gives a good account of the U.K. system of factory inspection of which the Health and Safety at Work Act is only the latest development.)

MYTH 52

The best way of conveying information to people is to tell them.

It is certainly true that talking is a more effective way of getting your message across to other people than writing to them, although we may have to follow up with the written word to reinforce, to explain details, and to provide a source for reference. Less important, the written word may provide evidence that we conveyed the message. Talking to people singly or in small groups is, of course, more effective than talking to large groups because feedback is easier—we know whether the message is received and what reaction it causes.

If we have to convey messages that people want to receive ("Where to get free beer," for example) almost all methods of communication are effective. However, if there is some resistance to our message, as there often is when we are making recommendations to increase safety, for example, then we should choose the most effective method of communication: discussion.

The following technique has been found particularly effective for putting across safety messages.

An accident is illustrated by a slide. The discussion leader explains very briefly what happened. The group then questions the discussion leader to establish the rest of the facts and say why they think the incident occurred. They then say what *they think* should be done to prevent the incident from happening again, not only in the plant where it actually occurred but in other plants as well.

Such discussions take longer than a lecture, but more is remembered and people are more committed to the conclusions, because they have not been told what to do but have worked it out for themselves.

If possible the incidents discussed should have occurred in the same company or factory, but if suitable incidents are not available sets of notes and slides can be obtained from the Institution of Chemical Engineers.[1]

The best size for the group is 12–20. If fewer than 12 are

present the group may not be 'critical' and discussion may not take off. If more than 20 are present the quieter members may not be able to contribute.

Similarly, if operating or safety instructions are being changed, do not just issue the new instructions. Instead, explain them to those who will have to carry them out; listen to their comments and answer their questions. You may find that what you propose is in some respects impracticable.

An argument against discussions is given in one of Harry Kemmelman's novels.[2] A group of students suggest a discussion instead of a lecture and the lecturer asks if they hope to achieve knowledge by combining ignorance.

However, in many situations people have the knowledge but do not seem able to draw on it. The stimulus of a discussion often helps them to do so, helps to break down the barriers between different parts of the mind. Also, while we cannot achieve knowledge by combining ignorance we may achieve knowledge by combining our individual bits of knowledge.

References

1. Hazard Workshop Modules. Published by the Institution of Chemical Engineers, 165–171, Railway Terrace, Rugby, CV21 3HQ, U.K.
   1. Hazards of Over- and Under-Pressuring of Vessels
   2. Hazards of Plant Modifications
   3. Fires and Explosions
   4. Preparation for Maintenance
   5. Furnace Fires and Explosions
   6. Human Error
2. Kemmelman, H. 1976. *Tuesday the Rabbi Saw Red.* p. 49. Greenwich, CT: Fawcett Publications.

MYTH 53

We can rely on the advice of the accountant where money is concerned.

Usually we can, but I describe below cases where the wrong technical decisions have been made as the result of following the company's accounting procedures.

1. Capital and Revenue

In many companies it was at one time the practice (in some it still is) to treat capital and revenue as if they were different commodities that could not be combined. As a result designs were chosen that were cheap in capital but expensive to maintain. The 1960s were the time of the 'minimum capital cost' philosophy. Since then industry has on the whole become more realistic—discounted cash flows have been widely used—but errors still occur.

For example:

1. A colleague of mine once watched a new plant being built from his office window. He calculated that by the time the plant was complete the cost of hiring the scaffolding around the distillation columns would have paid for a permanent structure, without taking into account the cost of hiring and erecting scaffolding for shutdowns during the life of the plant.

2. When equipment has to be isolated for maintenance the most effective method of isolation is a slip-plate (blind or spade) [Figure 53.1(a)]. A variation on the slip-plate is the figure-8 (spectacle) plate [Figure 53.1(b)]. In one position it allows flow; when it is turned it acts as a slip-plate.

   The disadvantage of the figure-8 plate is its extra cost. However, it is fixed in position. In contrast, slip-plates disappear when not in use and more are forever having to be ordered. Figure-8 plates may be cheaper in the long run, especially when isolation for repair is frequent.

3. Suppose that we wish to prevent a pump from over-heating if a valve in the delivery line is closed. Two methods are considered: installation of a high-temperature trip that will shut down the pump [Figure 53.2(a)] and installation of a kickback line from the pump delivery back to storage, so that there is always a small flow [Figure 53.2(b)]. Let us suppose that (a) costs $4000 and (b) $6000.

   At first sight it would seem that we should choose

(*a*) Slip—plate

(*b*) Spectacle plate

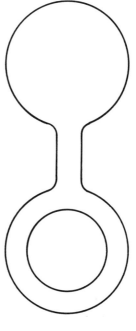

FIGURE 53.1

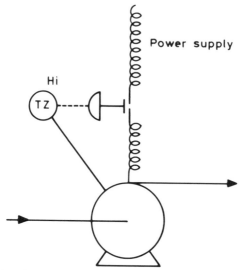

(a) High temperature trip

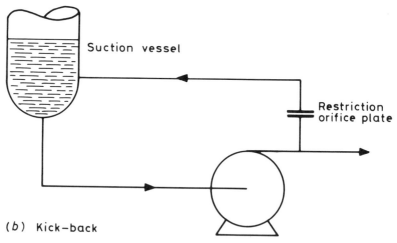

(b) Kick-back

FIGURE 53.2   Two methods of preventing a pump from overheating.

a, but the trip requires testing and maintaining and even after discounting this may double the capital cost. The pipeline, in contrast, has little maintenance costs attached to it, though it does use a little power as the pump continues to run. (In many real-life cases the pipeline would be cheaper in capital cost as well as revenue cost.)

Discounted cash flow (DCF) is widely recommended as a way of overcoming the dichotomy between capital and revenue but, as Malpas pointed out,[1] if we are not careful it can stifle innovation. When interest and thus discount rates are high, benefits that will not materialize for several years appear to have little value.

Malpas quoted an example:

Process A (established): Assume raw material costs steady for 3 years and then rise 10%/year.

Process B (new): Assume raw material costs initially double those for Process A but fall rapidly, becoming equal in 5 years and then rising at 5%/year.

After 6 years the cash flows are the same and from then on much in favor of B. But DCF calculations favor A.

What the DCF figures do show is that Process B is favored only from a long view.

2. Notional Capital

On a large integrated site a central organization supplied steam and other services to a large number of plants operated by other divisions of the same company. In this situation there are several possible ways of charging the user plants for services:

1. Market value—what it would cost the consuming plants to buy the services elsewhere (if they could) or manufacture them themselves.

2. Cost of production plus a reasonable return on capital (say 10–15%). Because there was a reasonably assured market for the services this return would be less than expected for a new product that carried a good deal of market risk.

3. The consuming plants carry a share of the capital of the services plants, proportionate to the amount of steam and other services they expect to use. There is a notional transfer of capital from the services organization to the user organization.

Method 3 was actually adopted. If the user plant was profitable and earned, say, 30% on capital, it had to pay this on services capital, and this made steam and other services very expensive.

A new plant expected to be a large user of steam, for driving compressors. The project team believed that they could not afford the high cost of the steam so they decided to install their own power source to drive the compressors. They chose free-piston gas generators. At this time (the 1950s) they were considered promising if not completely proven. Unfortunately they required extensive maintenance and were expensive to run.

If the services organization had been less "greedy" and had been content with a more modest return on capital the consuming organization would have used steam. The profits of the organization as a whole would have been greater.

3. Allocation of Overheads

In the Mexican branch of an international chemical company overheads were allocated in proportion to the number of invoices issued. This imposed a ludicrous penalty on a business (pharmaceuticals) that sold small quantities to large numbers of customers. It was found to be easier to transfer the business to an agent than to change the accounting system.

4. Costing Similar Products Together

Sir John Harvey-Jones, former chairman of ICI, has described a group of products that had performed well for many years[2]:

The cost accountancy criteria had been set up many years before, and had in their day been considered to be outstanding examples of a sensible way of allocating costs and overheads across an enormous range of associated products. The system appeared to work well for many years, but the business met hard times and many changes had to be made . . . we put in a new accountant from a totally different background, who examined the whole basis of our costing, *ab initio*. Within a remarkably short space of time we found, as is almost always the case, that what we thought was a universally bad business contained some products which were in reality extremely good, and a minority which were extremely poor. The minority had concealed the performance of the good but because we had looked at the whole lot together we had failed to appreciate this. This may seem a very elementary mistake for a large and sophisticated organisation to make, but I would caution against too happy a view that it could not happen in your organisation . . . ."

5. One Industry Makes the Products of Another

A final example is more concerned with national rather than company accounting. Methods have been developed for manufacturing single-cell protein from hydrocarbons or methanol.[3] The product is intended primarily for sale as a constituent of animal feedstuffs as an alternative to soya bean meal.

The raw materials for single-cell protein are the products of the oil and petrochemical industries, and manufacture has been carried out by oil and petrochemical companies, so the projects have been evaluated as typical oil and petrochemical projects and subjected to rigorous investment analysis. In most countries agricultural projects are assessed somewhat differently. Subsidies or guaranteed prices are often available to encourage domestic food production or even for sentimental reasons ("Farming is part of our heritage"). Should single-cell protein manufacture be assessed as a farming project or as a typical manufacturing project?

If new methods of making agricultural products are assessed as manufacturing projects, can they ever compete?

References
1. Malpas, R. 1983. *Harvard Business Rev*, July-August, p. 122.
2. Harvey-Jones, J.H. 1988. Making It Happen. London: Collins.

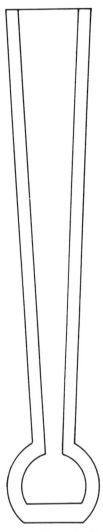

FIGURE 54.1   Do not try to debottleneck this bottle!

3.   Craig, J.B., and Lloyd, D.R. 1984. *ICI Technology for SCP Pro-duction from Methanol and its Wider Application.* Proceedings of the CHEM-RAWN III Conference (CHEMical Research Applied to World Needs), The Hague, Netherlands: Royal Dutch Chemical Society. June.

MYTH 54

Debottlenecking is the best way of increasing the capacity of a plant.

Debottlenecking often gives extra capacity for low capital cost. However, as Malpas pointed out,[1] there are disadvantages. Excessive debottlenecking draws scarce technical resources away from innovation and delays the date when brand new plants are built. It is these new plants, if they utilize new technology, that can make the greatest contribution to productivity, energy saving, environmental improvement, and safety.

The *extra* capacity obtained by debottlenecking may be bought cheaply but the effect on total costs is often small compared with the economies possible by introducing new technology. If output is limited by a few pumps or heat exchangers, obviously we replace or supplement them, but some bottles are nearly all neck (Figure 54.1).

Reference
1. Malpas, R. 1983. *Harvard Business Rev,* July-August, p. 122.

MYTH 55

I don't need to get involved in the detail.

Many a young engineer, after promotion, decides to stand back, take a broad view and leave the detail to his subordinates. There is no shortage of books on management recommending this practice. R.V. Jones puts the opposite view[1]:

> The experience brought home to me what my real strength had been at the earlier meetings. It was that, in contrast to everyone else sitting round the Cabinet table, I had done all my own work for myself and had forged out every link in the chain of evidence, so that I knew exactly what its strength was. Everyone else, in their more elevated positions, had had to be briefed at the last moment, as I myself had had

to be on this occasion. And even with Charies Frank's understanding and skill, and even though I had been away from the work for only a week, 1 felt that there was too much sloppiness in my knowledge for me to pronounce positively on the various possibilities. On a previous occasion I had quoted Palmerston's statement of 1838 to Queen Victoria, and now 1 even more appreciated its force:

"In England, the Ministers who are at the heads of the several departments of the State are liable any day and every day to defend themselves in Parliament; in order to do this they must be minutely acquainted with all the details of the business of their offices, and the only way of being constantly armed with such information is to conduct and direct those details themselves."

Professor Jones also describes how two physicists were assigned to work on the characteristics of German magnetic mines. One stuck to his desk, working on reports from minesweepers; the other went to sea and soon realized that the minesweepers' reports were hopelessly inaccurate.

We need to immerse ourselves in the detail—otherwise we do not understand what is happening—but we should then stand back and view the wider scene. This is not easy.

To quote Sir John Kendrew, "To turn out good scientists you have to keep their noses to the grindstone quite a lot, but you must at the same time try and stimulate their interests on a broad front, so that, when they do reach the stage of being independent creative people, they will go off into new fields; and this is very difficult."[2]

Summarizing the lessons of Three Mile Island, J.F. Ahearne wrote, "If the boss is not concerned about details, his subordinates will also not consider them important . . . . It is hard, monotonous, and onerous to pay attention to details; most managers would rather focus on lofty policy matters. But when details are ignored, the project fails . . . . Most managers avoid keeping up with details; instead they create 'management information systems.' "[3]

References

1. Jones, R.V. 1978. *Most Secret War*. p. 353. London: Hamilton.
2. Kendrew, J. 1974. *Chemistry in Britain* 10(11):443.
3. Ahearne, J.F. 1986. Three Mile Island and Bhopal: lessons learned and not learned. In *Hazards: Technology and Fairness*. p. 197. Washington, D.C.: National Academy Press.

MYTH 56

Spending money to prevent accidents will reduce the injury and damage they cause.

Usually it does, but not always. In the United States flood damage has increased as expenditure on flood control has increased.

The reason? When floods were frequent there was little development in areas subject to flooding. When flood prevention measures were introduced, development took place in these areas. When an exceptional flood overwhelmed the flood prevention measures (or the measures failed to achieve design performance) the damage was much greater than in the days before flood prevention was practiced.[1]

Efforts were made to reduce the risk of forest fires in the U.S. National Parks. They were at first successful. Fires were fewer and scrub and brushwood increased. When a fire did occur, it was of a size and intensity never experienced before.[1]

A tank was filled once a day with sufficient raw material to last for 24 hours. An operator watched the level and switched off the filling pump when the tank was nearly full. This took place without incident for 5 years until one day the operator allowed his attention to wander, and the tank was overfilled. The spillage was small because the operator soon saw it.

A high-level trip was then fitted to the tank. The designer's intention was that the operator would continue to watch the level and that the trip would take over on the odd occasion when the operator forgot to do so. The chance of the operator and trip failing at the same time was negligible. However, it did not work out like this. The operator left the control of the level to the trip and got on with other work. The trip failed, as could have been predicted, within 2 years, and the spillage was much greater than before because the operator was not on the job.

On a larger scale, one of the most important factors influencing the probability of death per year is social class. Accidents are small in comparison. In the United Kingdom today the mortality of adult males in social class 5 (unskilled occupations) is 1.8 times that of adult males in social class 1 (professional occupations). For children the ratio is slightly greater. Instead of taxing poor people so that the government can spend money on reducing remote risks, like those from nuclear power stations, would we save more lives if we left the money untaxed, or used it to increase the standard of living of the poor?[2]

References

1. Clark, W.C. 1980. In *Societal Risk Assessment.* p. 287, eds. R.C. Schwing and W.A. Albers. New York: Plenum Press.
2. Wildavsky, A. 1980. *The Public Interest,* Summer, p. 23.

MYTH 57

We need to know what's new.

Taken literally this is true, but there is an implication that we need be less concerned with what is old, and that is certainly not true. In my own subject of loss prevention and process safety, the majority of accidents have well-known causes. Occasionally an accident occurs because no one realized that A and B mixed together will react violently under certain conditions, but such accidents are the exception (and even here it is well established that tests should be carried out). Most accidents are very similar to accidents that have happened before, as the following examples show:

1. To start with a simple mechanical accident, a member of the U.K. Health and Safety Executive wrote:

    It is a chastening thought that, despite all the efforts of inspectors over the years and the accumulated experience of accidents, the belief is still current in some quarters that smooth rotating shafting is not dangerous, and accidents continue to happen on shafting as each generation re-learns the lessons of its predecessors."[1]

2. On plants that regularly use flexible hoses for transferring liquids or gases, men are often injured when removing hoses that still contain some liquid or gas under pressure. They may be injured by the liquid or by sudden movement of the hose. After each incident it is realized that the pipe to which the hose is connected should have been fitted with a vent valve so that the pressure can be released safely (Figure 57.1). There is a campaign for fitting vents. After a couple of years the accident is forgotten and recurs. A paper entitled "Accidents of the Coming Year"[2] describes a number of similar accidents that are regularly repeated.

3. More serious accidents recur after 10 or more years. People leave, the reasons for the precautions are forgotten, the precautions lapse, and the accident happens again. A paper entitled "Organizations have no Memory"[3] describes four fatal accidents that recurred after 10 or more years in the same organization. Here is one of them.

   In 1928 a 36-inch diameter gas main was being modified and a number of joints had been broken. The line was isolated from a gasholder by a closed isolation valve that, unknown to those concerned, was leaking. The leaking gas ignited. There was a loud explosion and flames appeared at various joints on the main. One man was killed.

   The source of ignition was a match struck by one of the workmen so that he could see what he was doing. However, once an explosive mixture is formed a source of ignition is always liable to turn up (see Myth 11). The real cause of the explosion was not the match but the leaking valve.

   The following are the conclusions of the original report:

   1. Never trust an open gas main that is attached to a system containing gas, and keep all naked lights clear.

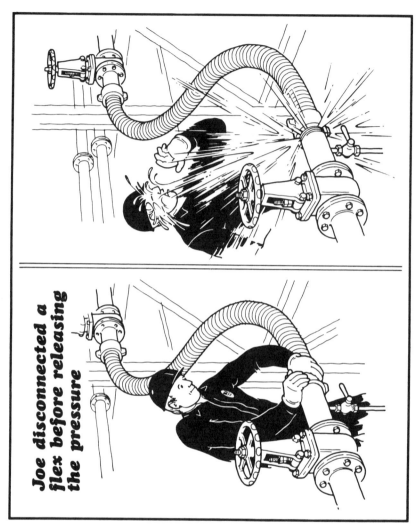

Joe disconnected a flex before releasing the pressure

FIGURE 57.1

138

2. When working on pipebridges at night, adequate lighting should be available.
3. Never place absolute reliance on a gas-holder valve, or any other gas valve for that matter. A slip-plate (blind or spade) is easy to insert and absolutely reliable.

In 1967, in another such incident, a large pump was being dismantled for repair. When a fitter removed a cover, hot oil came out and caught fire because the suction valve had been left open. No slip-plates had been fitted because it was not custom and practice to fit them. Three men were killed.

The oil was above its auto-ignition temperature. After the fire instructions were issued that before any equipment is given to maintenance:

1. The equipment must be isolated by slip-plates or physical disconnection unless the job to be done is so quick that fitting slip-plates (or physical disconnection) would take as long *and* be as hazardous as the main job.
2. Valves used to isolate equipment for maintenance, including isolation for slip-plating or physical disconnection, must be locked shut with a padlock and chain.
3. When there is a change of intent, for example, if it is decided to dismantle a pump and not just work on the bearings, the permit to work must be handed back and a new one taken out.

In the period of nearly 40 years that elapsed between these two incidents, the practice of slip-plating had lapsed—no one knew when or why. Perhaps the men who remembered the original incident had left and their successors did not see the need for slip-plating. "It is a lot of extra work," they might have said, "Other companies don't do it."

How can we keep old knowledge alive, especially details of past incidents and the action needed to prevent them happening again? The following suggestions may help:

1. Remind people of the incidents every few years in safety newsletters and other publications. These should recycle old knowledge as well as describe new.
2. Better still, hold regular discussions of past incidents, their causes, and prevention, as discussed under Myth 52, for people at all levels.
3. Devise better information storage and retrieval systems so that we can readily locate details of past incidents and recommendations on subjects on which we become interested.[4]
4. On each plant, start a black book containing reports of previous incidents of continuing interest. Do not clutter it up with reports on falls off bicycles and thumbs hit by hammers, but do include reports of interest from other companies.
5. Add a note to each standard and specification explaining *why* it has been adopted.
6. Design plants so that recommendations made after an accident can be carried out. (Slip-plating may have lapsed after 1928 because pipework was not designed with sufficient flexibility for slip-plates to be inserted.)
7. Remember that the first step down the road to the 1967 accident occurred when a manager turned a blind eye to a missing blind.
8. Spend less time reading magazines that tell us what's new and more time reading books that tell us what's old.

References

1. Watson, G.W. 1983. *In Her Majesty's Inspectors of Factories 1883–1983: Essays to Commemorate 150 Years of Health and Safety Inspection.* London: Her Majesty's Stationery Office.
2. Kletz, T.A. 1976. Accidents of the coming year. *Loss Prevention* 10:151.
3. Kletz, T.A. 1980. Organzations have no memory. *Loss Prevention* 13:1.

4. Fawcett, R.W., and Kletz, T.A. 1982. *Plant/Operations Prog.* 1(1):7.

MYTH 58

To prevent accidents at work we need to spend more money.

Sometimes we do; safety by design should always be our aim, but often there is no reasonably practicable way of removing a hazard by a change in design and instead we have to improve the software—that is, the method of working, the training, instructions, inspections, audits, and so on. We cannot spend our way out of every problem. About half the accidents that occur cannot be prevented by design changes, and a change in software is necessary.

Design changes are easy. All you need is money; if you make enough fuss you get that in the end and once you have got the new hardware it is unlikely to disappear. In contrast, changes to the software are subject to a form of corrosion more rapid than that which affects the steelwork. Procedures can vanish without trace in a few months once managers lose interest. A continuing effort, what I have called 'grey hairs,"[1] is necessary to maintain the software.

Furthermore, design changes need not cost money. If we make them early in design we can often *avoid* hazards and do not need to add on safety equipment to keep them under control (see Myth 26).

References
1. Kletz, T.A. 1984. *Plant/Operations Prog.* 3(4):210 and 5(1):J11.

MYTH 59

To prevent accidents we need to persuade reluctant managers and supervisors to take action.

Sometimes we do, but often it is just as necessary and more difficult to persuade foremen and operators to follow the

new procedures that managers and supervisors have introduced or use the new equipment that they have bought. Managers sometimes say, "It must be safe because we have done it
this way for 20 years without an accident" (see Myth 33).
Foremen and operators say this more often than managers,
and it is harder to convince them that the experience is irrele-
vant unless an accident in the 21st year is acceptable. The
manager, among his other skills, has to be an expert in the art
of persuasion. No wonder managers (and safety advisers)
have grey hair!

MYTH 60

Policies lead to actions.

The big men at the top lay down policy and the rest of us
carry it out. That is the theory. The practice is often different.
We solve our problems, as best we can, subject to various
pressures and constraints. Looking back, we see some consis-
tency in our actions. This is our policy. Statements of policy
are often a tidying-up operation: putting into statute form
what has become the common law of the organization. Actions
lead to policies, not policies to actions.

If you want to change the way something is done, say, the
way equipment is prepared for maintenance, you could start
by persuading the directors to issue a statement of policy.
After a serious accident people are willing to accept direction
and this may be the right solution. At other times it might be
better to persuade individual managers—the lowest level
that has the power to change the procedures. When most of
them are operating the new procedures, that is the time to
issue a new policy statement.

Index

Page numbers in italics indicate figures; Page numbers followed by t indicate tables.